GEOMETRICAL DRAWING

by

G. PEARSON

St. John's College, York

THIRD EDITION (metric)

OXFORD UNIVERSITY PRESS 1972

Oxford University Press, Ely House, London W. 1

GLASGOW NEW YORK TORONTO MELBOURNE WELLINGTON
CAPE TOWN IBADAN NAIROBI DAR ES SALAAM LUSAKA ADDIS ABABA
DELHI BOMBAY CALCUTTA MADRAS KARACHI LAHORE DACCA
KUALA LUMPUR SINGAPORE HONG KONG TOKYO

Printed in Great Britain by
Fletcher and Sons. Ltd. Norwich

PREFACE

The aim of this book is to provide a course in Geometrical Drawing which will lead to work at and beyond 'O' level of the General Certificate of Education. The book is not confined entirely to the General Certificate, because it covers the basic work which is essential before a real attempt can be made to introduce Machine, Architectural, or Building Drawing. In fact, having done a course in Geometrical Drawing, a draughtsman is able to concentrate on the subject of his drawing, e.g. a building or a machine part, rather than on drawing technique.

I believe that in learning Geometrical Drawing it is much better to depend upon a reasoned rather than a mechanical approach to the subject; to put it more simply, to build upon what has already been learnt in simple geometry. I have dealt with the subject—so far as it allows—by reasoned explanation of both the construction and the future application. The latter I regard as extremely important, because it shows that there is purpose in doing this type of drawing. The course is intended to be a continuous one in which one problem leads naturally to the next, ensuring that the basic principles have to be learnt and can be easily used in future work. The book is divided into two sections, Plane and Solid Geometrical Drawing, and at the end of each section are test papers consisting of questions set in the General Certificate of Education.

Solid Geometrical Drawing largely depends upon a knowledge of Plane Geometrical Drawing, hence this latter and more abstract section is placed first. In the Solid Geometrical Drawing, I am convinced that the approach must be concrete rather than abstract, so I have suggested the use of a pin to represent a point and line, a piece of tinplate or card to represent a lamina, various shaped wooden solids, and, most important of all, the constant use of paper planes.

I make no apologies for my method of introducing orthographic projection. It is not illogical to start, after explaining what is meant by orthographic projection, to draw the projections of a point and then a line. This is a means of introducing planes and the simple traces of a line, and forms an introduction to the more difficult chapter on the traces of planes.

I regard the chapter on the traces of planes as vital to all future work. Every view is projected on to a plane, and the work of subsequent chapters depends on the use of imaginary planes, either cutting planes or auxiliary planes, by means of which true shapes, sections, and auxiliary views are drawn.

I have placed isometric and oblique projections at the end of the book, not because I think that they should not be taught earlier, but rather to preserve the continuity of the chapters on orthographic projection. In fact it is a good idea to introduce isometric and oblique projections at an earlier stage and to use them as a break from the more difficult work on orthographic projection. There is no doubt that discretion will have to be exercised, both in introducing isometric and oblique projection, and also in omitting parts which are not required for some of the G.C.E. syllabuses.

Most authors rely on the assistance of different people and bodies in order to make improvements and remove imperfections before publication. I do not claim to be an exception to this, and my sincere thanks are due to those colleagues and boys of the Lawrence Sheriff School who have helped in various ways, and particularly to my colleague, Mr. D.V. Skeet, for his invaluable work in reading, correcting, and suggesting improvements.

I gratefully acknowledge the kindness of the examining bodies in granting me permission to reprint General Certificate of Education questions: the Joint Matriculation Board in Geometrical Drawing (taken from Geometrical and Engineering Drawing and Building Construction and Geometrical Drawing); the Oxford Local Examinations in Technical Drawing; the Oxford and Cambridge Schools Examination Board in Geometrical and Mechanical Drawing; the University of Cambridge Local Examinations Syndicate in Technical Drawing; the University of London in Geometrical and Mechanical Drawing; and the Welsh Joint Education Committee in Technical Drawing.

G.P.

Rugby
March 1956

PREFACE TO THE SECOND EDITION

In this revised edition the first object has been to make any necessary corrections. Some changes in construction have been made, in particular the diagonal scale and cycloid, the former to emphasize that a diagonal scale consists of a number of plain scales and the latter to serve as the basis of other constructions of cycloidal curves. Mention has been made of sloping letters and figures, also that oblique projection can be drawn at angles other than 45°. The general locus construction of conic curves and some exercises involving the practical application of plane geometry have been added.

The General Certificate questions have been replaced with some of more recent date, and I gratefully acknowledge the kindness of the examining bodies in granting me permission to reprint their questions. The examining bodies are those listed above in addition to the Associated Examining Board in Geometrical Drawing (Building and Engineering); and the Southern Universities' Joint Board for School Examinations in Geometrical and Machine Drawing.

G.P.

York
January 1967

PREFACE TO THE THIRD (METRIC) EDITION

The main object in this revision has been to change the imperial measurements to metric and to ensure that all the drawings are in accordance with the present British Standard 308: *Engineering Drawing Practice.*

All questions have also been changed to metric units with the permission of the respective examination boards, though the author accepts responsibility for any errors that might have resulted from this change. In some instances, General Certificate Questions have been replaced by others. The author gratefully acknowledges the permission of the following boards to reprint questions taken from recent papers: the Joint Matriculation Board (metric papers); the Oxford Local Examinations and the Oxford and Cambridge Schools Examination Board.

G.P.

York
July 1972

CONTENTS

SOLID GEOMETRICAL DRAWING

CHAPTER 1

INTRODUCTION

Machine drawing is a universal language in which the designer's ideas are transferred to the work bench by means of an accurate drawing. Because most countries have their own Machine Drawing standards, the method of writing and interpreting this language varies, mainly in the angle in which the drawing is made; for example, in America the third angle is used, in Britain the first and third angles.

To produce such a drawing it is essential that the draughtsman should be able to make a scale, draw different kinds of figures, represent a solid by views on a plane surface, make dimensioned sketches, and follow a definite standard method of drawing.

Plane and Solid Geometrical Drawing are the basic principles of Machine Drawing and to produce them accurately and legibly—because they are made to be read—good paper, good-quality instruments correctly used, and logical reasoning are essentials. Without this last the others are useless.

Cartridge paper of A1 (594 X 841mm), or fractions of it, is normally used for pencil drawings. The most common size used in schools is A2

Set Squares. These are used with a tee square to draw vertical or inclined lines.

Scale Ruler. This is used for measuring directly a scaled dimension.

Dividers. These are used to transfer measurements from the scale ruler to the drawing.

Protractor. This is used for measuring or drawing angles.

Compasses. These are used to describe circles and arcs.

Pencil. This is the most important piece of equipment, because pencilwork will either make or mar a drawing. Therefore two essentials must be followed:

(i) The pencil must be of the correct grade—grades vary from 6B to 6H, i.e. very soft to very hard—and, at this early stage, an H is recommended for both pencil and compass lead as this will produce a clean sharp line without cutting or blurring the paper.

(ii) It must be sharp.

For drawing lines either a chisel edge or a conical point can be used. The chisel edge (Fig. 1.1) gives a thin line and maximum life between sharpenings, but cannot be used for lettering. The conical point (Fig. 1.2) can be used both for drawing lines and lettering.

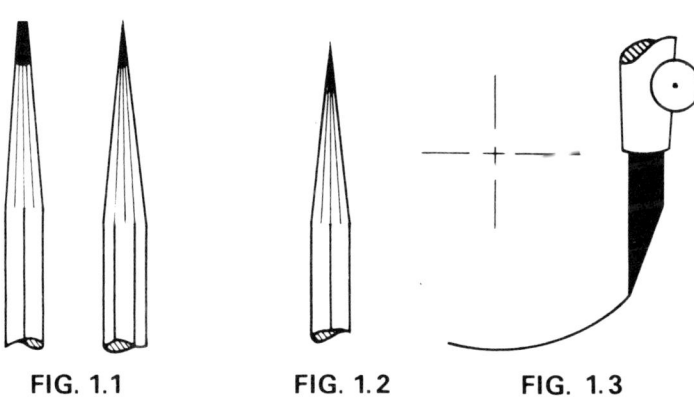

FIG. 1.1　　**FIG. 1.2**　　**FIG. 1.3**

(420 X 594mm). A grade H pencil is recommended as the most suitable for the early stages of Geometrical Drawing.

Equipment and Instruments

Drawing Board. This has a smooth flat surface and straight edges.

Drawing Clips. These are used to fasten the paper to the board.

Tee Square. This is used for drawing horizontal lines or as a horizontal edge for set squares.

Fig. 1.3 shows the shape of a compass lead if a chisel edge is used. It is most important that the grade of the compass lead should be correct, i.e. generally one grade softer than that of the pencil used, otherwise there will be a difference in density at the point where straight and curved lines join.

No drawing equipment is complete without a clean duster to dust both paper and instruments, and a soft rubber to remove superflous construction lines.

EXERCISES

1. Place the paper centrally on the board, with the top edge parallel to the top edge of the tee square. Fasten the paper at the four corners. With the flat side of the pencil against the tee square edge, draw two horizontal lines, 10mm from the top and 10mm from the bottom edges of the paper. Complete a 10mm margin round the paper by drawing two vertical lines.

2. Draw one horizontal line, 30mm below the top margin, and one vertical, 30mm from the left-hand margin.

3. Measuring from the intersection of the horizontal and vertical lines of Exercise 2, mark off seven points on the vertical each 26mm apart and draw seven horizontal lines through these points, each 240mm long.

4. Measuring from the intersection of the horizontal and vertical lines of Exercise 2, mark off the points on the horizontal each 22mm apart and through these draw ten vertical lines, each 225mm long.

5. From one centre describe four circles of 30mm, 60mm, 70mm and 90mm diameter.

6. By means of set squares, from one point draw angles of 30°, 45°, and 60°.

7. By means of a protractor and from a point on a straight line draw angles of 5°, 10°, 25°, and 50°.

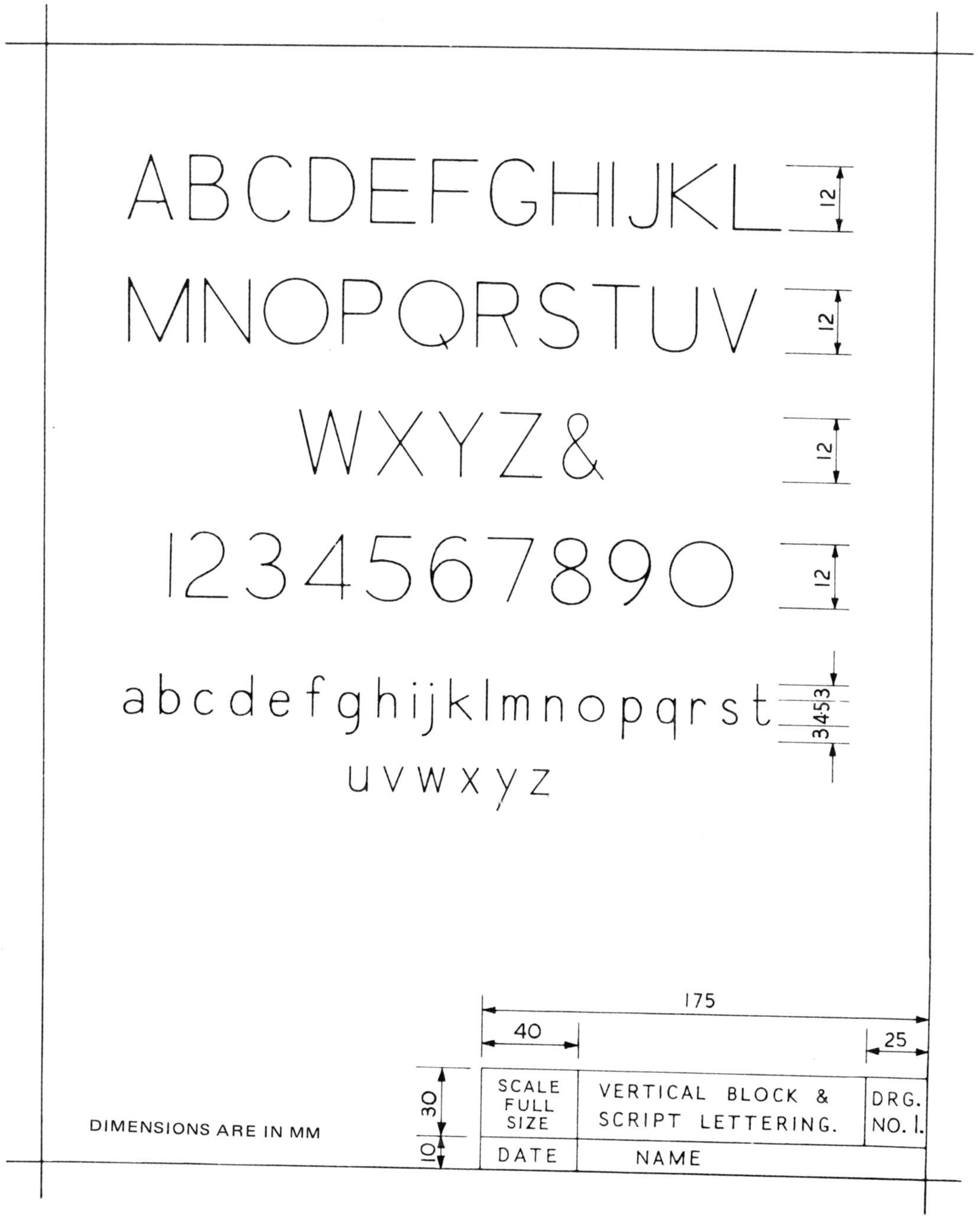

ABCDEFGHIJKL 12

MNOPQRSTUV 12

WXYZ& 12

1234567890 12

abcdefghijklmnopqrst 3 4 5 3

uvwxyz

DIMENSIONS ARE IN MM

175

40

25

30

10

	SCALE FULL SIZE	VERTICAL BLOCK & SCRIPT LETTERING.	DRG. NO. 1.
	DATE	NAME	

CHAPTER 2

LETTERING AND LAYOUT OF A DRAWING

Before a drawing is complete it is necessary to put on certain information in such positions that it can be easily read. This is done by Block or Script Lettering. Letters and figures can be either sloping or vertical, but since most characters are vertical—British Standards suggest the use of vertical characters for drawing numbers, titles, and reference numbers, i.e. in title blocks—it is much better to use only these. The difference between the two types of lettering is that the sloping characters are at an angle of approximately 20° to the right of the vertical, and it is a simple matter to form these once the production of vertical letters and figures is known.

Block letters are used for titles, name, scale, drawing number, and date. Script letters are used for detailed information relating mainly to processes or constructions, and both types of lettering must be capable of rapid reproduction in pencil or ink at a later stage and also made legible.

So that the drawing can be easily read and the essentials seen almost at a glance, some form of standard layout must be used. This consists of a margin to form the drawing boundary, good spacing of views, and a Title Block which gives general information. The drawing number is an easy means of quickly comparing early and later work.

Metric Units

Dimensions in metric units are usually expressed in millimetres and the symbol mm is omitted on drawings.

A note is placed on the drawing stating the units, as shown on page 3.

All units in this book, other than degrees, are expressed in millimetres.

Fractions

Fractions, with other figures or letters, are approximately 1¼ times the height of the figures or letters.

The numerator and denominator are equal in height with small spaces between each and the line.

Layout

In British Standards, layout is mainly the drawing of a frame, title block and the lettering used to give drawing numbers, part numbers, and other necessary information; however, in early work, particularly in schools, the correct spacing of drawings (or views) within a frame is considered the most important aspect, since a drawing is made to be read.

A suggested title block is shown on page 3, and a 10mm margin is recommended for A2 size paper.

Vertical Letters and Figures

Spaces between letters and figures must be equal.

Spaces between words must be equal.

There must be spaces between lines.

All letters and figures must be drawn between parallel guide lines.

Block Letters and Figures

O and Q are circles.

M and W fit into squares.

I is a vertical line.

J is *half* as wide as its height.

The remaining letters are *three-quarters* as wide as their height.

Script Letters

The widths have the same ratio as block letters.

The top and bottom spaces between the guide lines are each *two-thirds* of the middle space.

EXERCISES

1. Draw, (i) 10mm margin; (ii) the alphabet and figures 0—9 in block 12mm high; (iii) the alphabet in script, middle space 4·5mm high.

2. Draw a title block in the bottom right-hand corner of the paper. In block letters insert title, name, date, scale, and drawing number. Use your own discretion for the height of these letters.

3. Using figures 6mm high, show the following fractions: $1\frac{1}{2}$, $4\frac{1}{4}$, $\frac{1}{16}$, $2\frac{5}{16}$, $\frac{7}{32}$, $5\frac{13}{32}$.

4. First in block and then in script, letter the following: Horizontal and Vertical Planes.

PLANE GEOMETRICAL DRAWING

In learning a language the words are of first importance, and in learning Machine Drawing, Plane Geometrical Constructions are the words which, once known, can be translated into the views of small machine parts or huge castings.

CHAPTER 3

CONSTRUCTION OF LINES AND ANGLES

It is essential to know the construction and division of lines, angles, and circles, because these form an integral part of the drawing of plane figures.

FIG. 3.1 *To bisect (divide into two equal parts) a given line AB.*

Draw the line *AB*. With centres *A* and *B* and a radius greater than half *AB*, describe four arcs to intersect at *C* and *D*. Since the radius is the same for all four arcs, then *C* and *D* are equal distances from *A* and *B*, and a line joining *CD* must bisect *AB*.

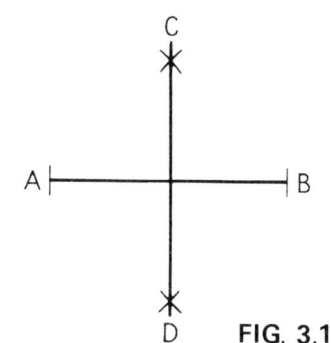

FIG. 3.1

FIG. 3.2 *To draw a perpendicular (or construct a right angle) from a given point P on a straight line.*

Draw the straight line. With *P* as centre and any convenient radius describe a semicircle to terminate at *A* and *B*. *P* is now the centre point of line *AB*. With *A* and *B* as centres and a radius greater than half *AB*, describe two arcs to intersect at *C*; as in Fig. 3.2, *CP* will be the bisector of the line *AB*, perpendicular (or at right angles) to *AB*.

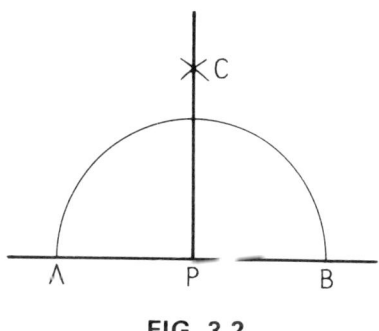

FIG. 3.2

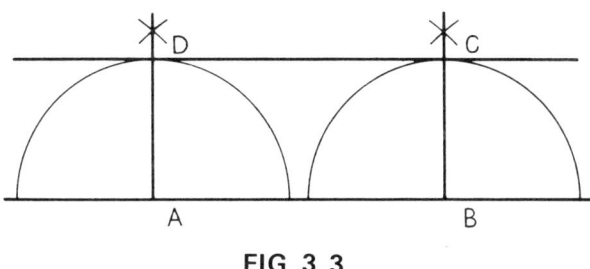

FIG. 3.3

FIG. 3.3 *To draw a parallel line a given perpendicular distance from a given straight line AB.*

Draw the straight line *AB*. With *A* and *B* as centres and a radius equal to the perpendicular distance, describe two semicircles. Construct perpendiculars at *A* and *B* to cut the semicircles at *C* and *D* respectively. Join *CD* (the common tangent) which is parallel to *AB* and is the required line.

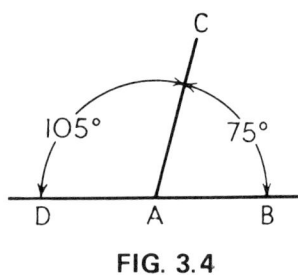

FIG. 3.4

FIG. 3.4 *To construct angles of 75° and 105°.*

Draw *AB* and produce to *D*. Since 45° and 30° equal 75°, then with the 45° set square on the tee square, the 30° set square on the 45° and one edge passing through *A*, draw *AC*. The 75° angle is *CAB*. Angle *DAB* is equal to 180°, therefore the supplementary angle *DAC* is 105°.

Angles can also be constructed by a Scale of Chords: this is based on the lengths of the chords of different angles measured on the same arc. It is a compass method of construction.

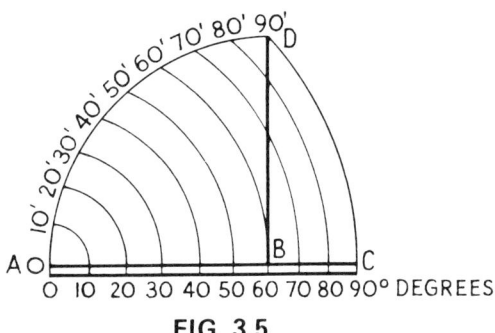

FIG. 3.5

FIG. 3.5 *To construct a scale of chords, reading 0° to 90° in 10°, given an arc of radius AB.*

Draw the straight line *AB* and produce. Erect a perpendicular at *B*, and with *B* as centre describe the arc *AD*. *AD* is the arc of the right angle *ABD*, therefore (by means of a protractor or dividers) divide the arc into nine equal parts; then by the fact that an angle is an amount of turning, each part will present 10° about the centre *B*. To construct the scale, the chords to each point of the arc have to be transferred to a straight line; then with centre *A*, radius chord *A*10′, describe an arc to cut *AB* produced at 10. From the same centre *A*, describe arcs of radii chords *A*20′, *A*30′, *A*40′, *A*50′, *A*60′, *A*70′, *A*80′, and *A*90′ to cut *AB* produced at 20, 30, 40, 50, 60, 70, 80, and 90 (point *C*). To complete the scale draw the rectangular strip about *ABC*.

The construction of an angle of 60° is a special case in which the length of the chord is equal to the radius of the arc on which it is measured.

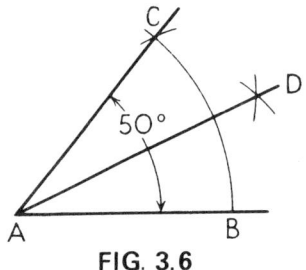

FIG. 3.6

FIG. 3.6 *To construct an angle of 50°, by a scale of chords, and to bisect the angle.*

Draw a straight line *AB*, making *AB* equal to *A*60°. With *A* as centre and radius *A*60°, describe an arc. With *B* as centre, radius chord distance *A*50°, cut the first arc at *C*. Join *AC* to complete the angle of 50°. Since the arc *BC* represents the amount of turning, the bisection of its chord will be the bisection of the angle, therefore with centres *B* and *C* and a radius greater than half *BC*, describe two arcs to intersect at *D*. Join *AD*.

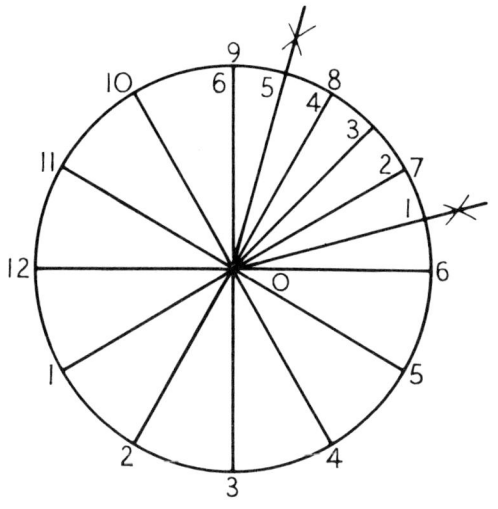

FIG. 3.7

FIG. 3.7 *To divide a circle, radius OA, into twelve and twenty-four equal parts.*

With centre O and radius OA, describe a circle. Through centre O divide the circle into angles of 30° by a 60° set square, this gives twelve angles of 30° and twelve equal parts. To divide the circle into twenty-four equal parts, bisect each angle of 30° at the centre and produce the bisectors to make six additional diameters. Only a quadrant is shown divided a second time. This is often termed 'radial division' of a circle.

EXERCISES

1. Bisect the following straight lines: 90mm, 75mm, 100mm, 103mm.

2. On a straight line 120mm long, erect a perpendicular 35mm from one end. Repeat for a 105mm line, the point 30mm from end.

3. Construct parallel lines: (i) 45mm above a straight line; (ii) 65mm above a straight line; (iii) 56mm above a straight line.

4. Construct angles of 75° and 105°.

5. (i) Construct a scale of chords, radius of arc 100mm, to read 0° to 90° in 5° intervals.

(ii) From this construct angles of 15°, 30°, 45°, 50°, 70°, and 85°.

(iii) Bisect the angles constructed in (ii).

6. (i) Divide a 100mm diameter circle into equal parts.

(ii) Divide a 100mm diameter circle into twenty-four parts.

CHAPTER 4

THE DIVISION OF STRAIGHT LINES AND THE CONSTRUCTION OF SCALES

In Machine Drawing it is sometimes necessary to construct a scale, and it is also of value to know how the scale ruler has been calibrated. The basis of this construction is shown in Fig. 4.1. The triangles ABC and $AB'C'$ are similar and by measurement it can be shown that

$$\frac{\text{side } AB}{\text{side } AB'} = \frac{\text{side } BC}{\text{side } B'C'} ;$$

Therefore any line can be divided in a given ratio or number of equal parts by constructing similar triangles upon the line. This method is more accurate and better than marking points off a ruler.

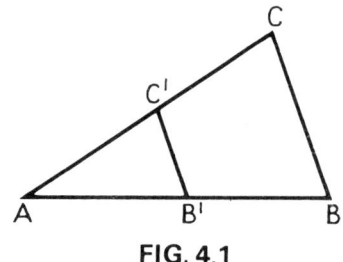

FIG. 4.1

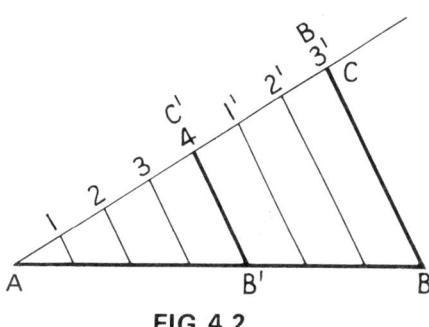

FIG. 4.2

FIG. 4.2 *To divide a line AB in a ratio of* 4:3.

Draw AB. From A draw AC of a convenient length and angle to AB. Step off seven equal distances from A on AC and number the points. Join the last point $3'$ to B and draw $4B'$ parallel to $3'B$; therefore two similar triangles have been constructed and $AB' : B'B$ equals 4:3.

To divide a line AB into seven equal parts.

Draw AB, AC and seven divisions on AC as in the ratio construction. Join $3'B$, i.e. the last point on AC to B. Draw parallels to $3'B$ from the other six points on AC, thus constructing seven similar triangles and dividing AB into seven equal parts.

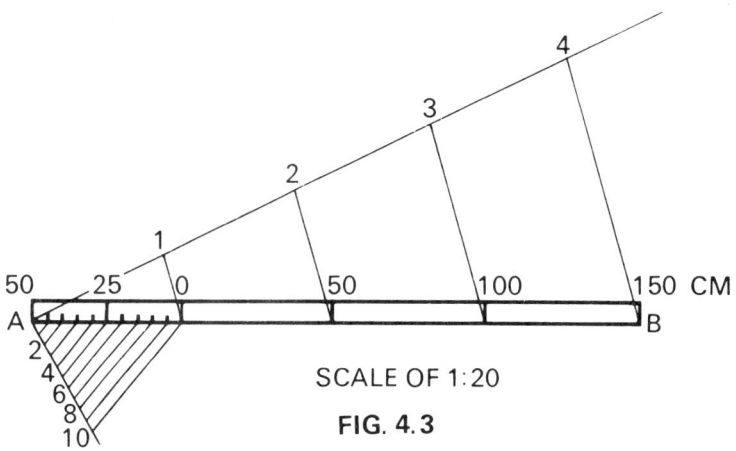

SCALE OF 1:20

FIG. 4.3

FIG. 4.3 *To construct a Plain Scale of* 1:20, *and to read up to* 200cm *in* 5cm *intervals.*

Draw AB 100mm long. Using the same construction as in Fig. 4.2, divide AB into four equal parts, and underneath the line AB divide the first division into ten equal parts. Draw a rectangular strip on AB and, from the points dividing AB, draw perpendiculars. Number and letter the scale. The position of O makes it possible to commence measuring from a full unit on the scale.

FIG. 4.4 *Principle of a Diagonal Scale.*

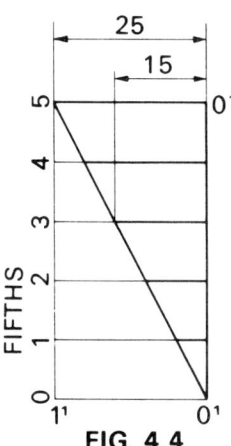

FIG. 4.4

Draw a vertical line and mark point 0 at the lower end. From 0 step off (with dividers) 5 equally spaced points and number 1, 2, 3, 4, and 5. Draw 0 0′ horizontal and 25mm long. From 0′ draw a vertical line and horizontal lines to meet it from the points 1, 2, 3, 4, and 5. Draw the diagonal 0′5.

The triangle 0′0′5 is now almost the same as that of Fig. 4.2, but 0′0′ (which represents *AB*) is vertical and angle 0′0′5 is 90°, not acute. The spaces on 0′0′ are equal, the spaces on 0′5 (which represents *AB*) are also equal and, since there are 5 spaces, the length of each horizontal line is $\frac{1}{5}$ less than that of the one above. Line 1 is 5mm and the others 10mm, 15mm, 20mm, and 25mm respectively. It is very difficult to draw several small lengths, such as 0·2mm on one line, but it is possible on a diagonal scale because it consists of several plain scales, which, when measured to a diagonal, differ by a small amount, in Fig 4.4 by 5mm.

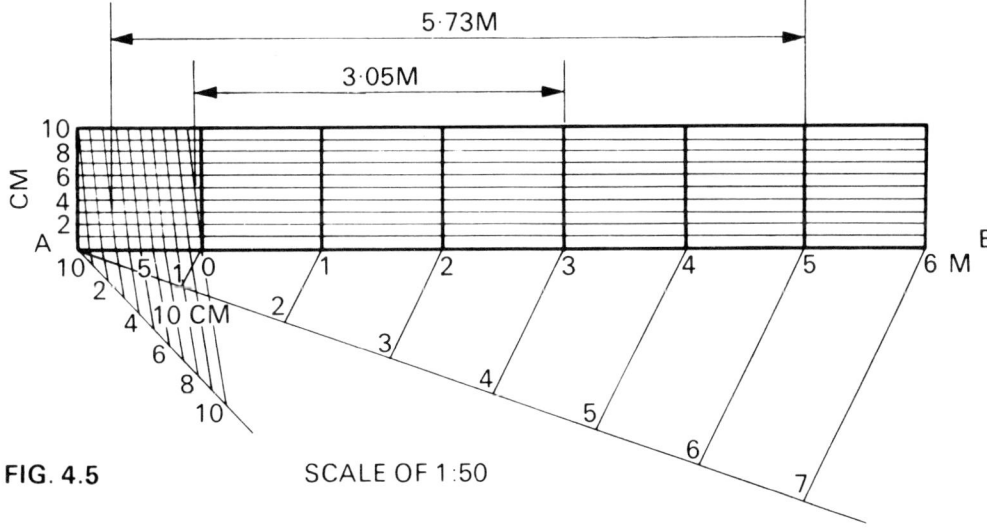

FIG. 4.5   SCALE OF 1:50

FIG. 4.5 *Construction of a diagonal scale of* 1:50 *to read up to 7m in metres and centimetres.*

Draw a straight line *AB* 140mm long, i.e. equal to $\frac{1}{50}$ of 7 metres. Divide *AB* into seven equal parts and the first part into ten equal parts. Erect a perpendicular at *A* and from *A* step off ten equally spaced points. Draw parallel lines to *AB* through each of these points. Erect perpendiculars on *AB* at 0 and, to the right of 0, through 1, 2, 3, 4, 5 and 6. Using the method shown in Fig. 4.4, draw the diagonal from the first division on *AB* to 10 on the perpendicular. Through the remaining points draw the other diagonals parallel to the first. Letter and number the scale.

To use the scale for measuring (i) 3·05m, (ii) 5·73m.

(i) From the perpendicular at 3 and along the parallel at 5, measure to the diagonal numbered 0.

(ii) From the perpendicular at 5 and along the parallel at 7, measure to the diagonal at 3.

Scales are sometimes referred to by their Representative Fraction (R.F.), which is the *actual measurement on the drawing compared with the real measurement of the object.*

EXERCISES

1. Divide a 135mm line in a ratio of: (i) 2:3, (ii) 3:4, (iii) 5:3.

2. Divide a 118mm line in proportion of: (i) 1:2:3, (ii) 1:3:4, (iii) 1:2:3:4.

3. Construct a plain scale 1:5, to read up to 1m in centimetres.

4. A line 115mm long represents a distance of 2m. Construct a scale on this line to read in metres and 5 centimetres. Show 1m 35 cm.

5. Construct a diagonal scale full size to read up to 15cm in centimetres, millimetres and 0·5 millimetres. Show 14·35cm.

6. The R.F. of a diagonal scale is 1:50. Construct the scale to read up to 6m in metres and centimetres. Show 3·75m.

7. A line 12cm long represents 5m. Construct a scale on this line to read in metres and centimetres. Show 3·07m.

8. A machine part, 150mm long, is dimensioned in millimetres and 0·2mm. Draw a scale from which these could be taken and indicate 76·4mm and 101·8mm.

CHAPTER 5

CONSTRUCTION OF TRIANGLES AND QUADRILATERALS

Triangles

A triangle is a common plane figure which can be the section of a roof or the V in a vee block, and, as such, its construction and various forms have to be known. It is bounded by three straight lines and is named with reference to either the sides or the angles.

Equilateral. All angles and sides are equal.
Isosceles. Two sides and two angles are equal.
Scalene. Sides and angles are *unequal*.
Right-angled. Has one right angle.

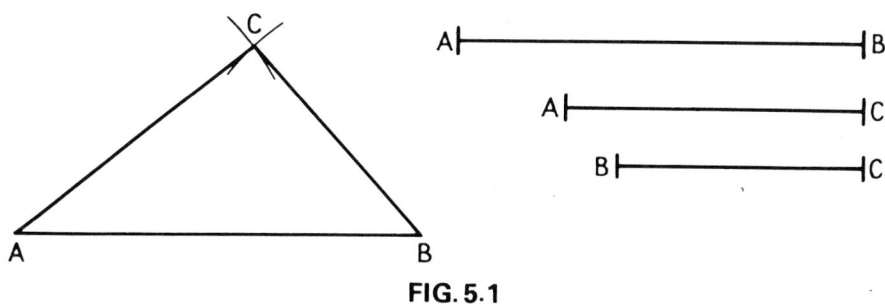

FIG. 5.1

FIG. 5.1 *To construct a triangle given the length of each side, AB, BC, and AC.*

Draw the side *AB*. With centre *A* and radius side *AC* describe an arc, then with centre *B* and radius side *BC* describe a second arc. The point *C* will be on both these arcs, therefore the intersection of the two arcs will be *C*. Join *AC* and *BC* to complete the triangle.

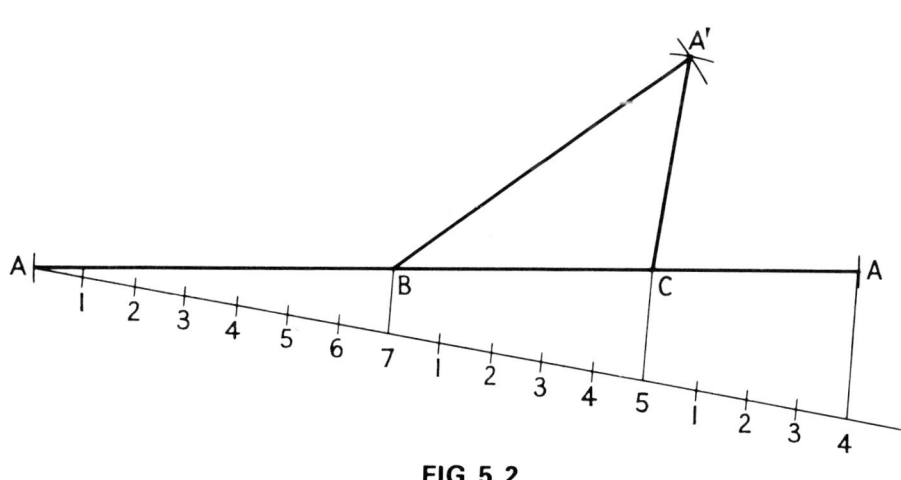

FIG. 5. 2

FIG. 5.2 *To construct a triangle given its perimeter AA, and the sides in proportion of* 7:5:4.

Draw *AA*, and by the method of Chapter 4, divide it into a proportion of 7:5:4, giving points *B* and *C*. The three sides are *AB, BC, and AC.* Using the method of Fig. 5.1, and with *B* and *C* as centres and *AB* and *AC* as radii respectively, draw two arcs the intersection of which will be point *A'*. Join *A'B* and *A'C* to complete the triangle.

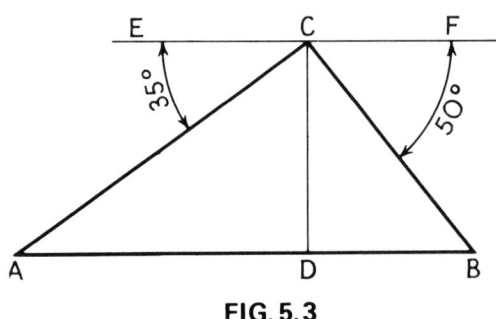

FIG. 5.3

FIG. 5.3 *To construct a triangle given the altitude CD, base angles of 35° and 50°.*

Draw the altitude *CD*. The triangle will be between two parallel lines perpendicular to *CD*. Construct a perpendicular to *CD* passing through *D* and produced in both directions. Construct a parallel line to this perpendicular passing through *C* (line *ECF*). Draw an angle of 35° on *EC* and produce to *A*. Draw an angle of 50° and produce to *B*. Since *AC* is a line across two parallels, angle *CAD* will equal angle *ACE*; therefore angle *CAD* is 35° and angle *DBC* equals 50° by the same reasoning.

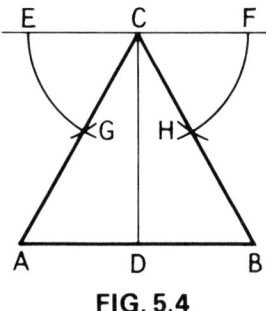

FIG. 5.4

FIG. 5.4 *To construct an equilateral triangle given the altitude CD.*

This is a special case of the method of Fig. 5.3, which only differs by the construction of the angles. Draw the altitude *CD* and draw perpendiculars to *CD* through *C* and *D*. With *C* as centre construct two angles of 60° by the chord method, giving *CG* and *CH*. Produce *CG* to *A* and *CH* to *B* to complete the triangle.

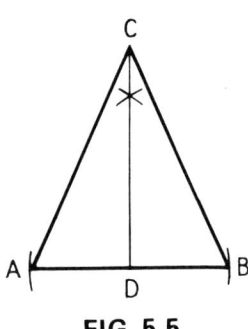

FIG. 5.5

FIG. 5.5 *To construct an isosceles triangle given the base AB and the altitude CD.*

Draw the base *AB*. Since the base angles are equal, then the altitude will bisect the base, therefore bisect *AB* at *D* and draw *CD* perpendicular to *AB*. Join *AC* and *BC* to complete the triangle.

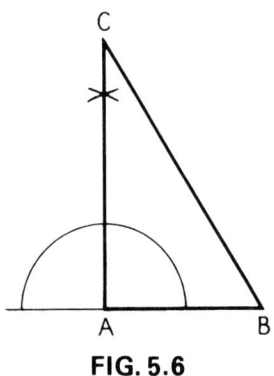

FIG. 5.6

FIG. 5.6 *To construct a right-angled triangle given the base AB and a side AC.*

Draw base *AB* and produce so that the perpendicular (the side *AC*) can be constructed at *A*. Join *BC* to complete the triangle.

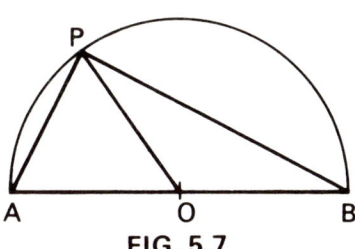

FIG. 5.7

FIG. 5.7 *To show the principle upon which a right-angled triangle can be drawn within a semicircle.*

Fig. 5.7 is that of a semicircle of centre O and diameter AB. P is any point on the circumference and AOP and BOP two triangles, together making the triangle APB. Since OP, OA, and OB are radii of the same semicircle and therefore equal, then the triangles are isosceles and so angle OAP is equal to angle OPA, angle OBP is equal to angle OPB. By addition, angle OAP + angle OBP is equal to angle APB (i.e. angles OPA + OPB). But these are the angles of the triangle APB and together equal 180°. Therefore angle APB is equal to 90° and the triangle is right-angled with its hypotenuse the diameter AB. By this method, if the hypotenuse and one side is given, it is possible to draw any right-angled triangle in a semi-circle.

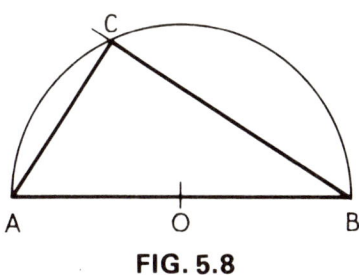

FIG. 5.8

FIG. 5.8 *To construct a right-angled triangle given the length of one side and the hypotenuse AB.*

Draw the hypotenuse AB and on it describe the semicircle. With A as centre and side AC as radius, describe an arc to cut the circumference of the semi-circle at C. Join BC and AC. The right angle is ACB.

Quadrilaterals

Quadrilaterals are plane figures bounded by four straight lines, and include the rectangle and square; these are often the shapes of all or parts of objects.

Square. All its sides are equal and its angles are right angles.

Rectangle. Its opposite sides are equal and all its angles are right angles.

Rhombus. All its sides are equal and its opposite angles are equal but not right angles.

Parallelogram. Its opposite sides are equal and its opposite angles are equal but not right angles.

Irregular Quadrilaterals. An irregular quadrilateral with two sides parallel is sometimes termed a trapezium.

FIG. 5.9 *To construct a square given the length of one side AB.*

Draw side AB, produce and erect a perpendicular at A to AB. With centre A and radius side AB, describe an arc to cut the perpendicular at D. With D as centre and radius side AB, describe an arc on which will be the point C. With B as centre describe a third arc of the same radius, this will cut the one from centre D at C. Join BC and CD to complete the square.

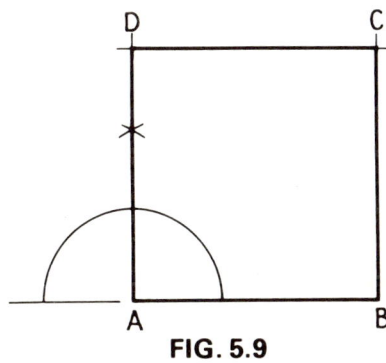

FIG. 5.9

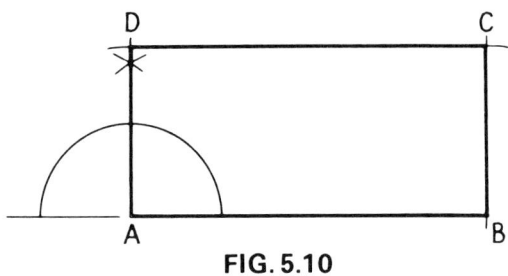

FIG. 5.10

FIG. 5.10 *To construct a rectangle given the length of two sides AB and AD.*

Draw side *AB* and produce to erect a perpendicular to *AB* at *A*. With *A* as centre and radius side *AD* describe an arc to cut the perpendicular at *D*. With *B* as centre and radius side *AD*, describe an arc, on which will be the point *C*, then with *D* as centre and radius side *AB*, describe an arc which will cut the one from centre *B* at *C*. Join *BC* and *CD* to complete the rectangle.

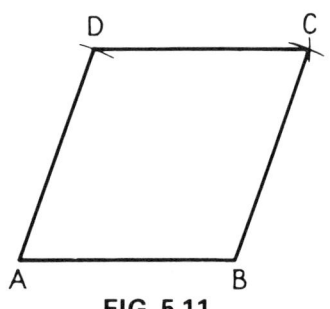

FIG. 5.11

FIG. 5.11 *To construct a rhombus given the length of one side AB, and an angle ABC.*

Draw side *AB* and angle *ABC*. Since the rhombus has four equal sides the remainder of the construction is that of a square, therefore with centre *B* and radius side *AB*, describe an arc to cut *BC* at *C*. With centres *A* and *C* and radius side *AB*, describe arcs to intersect at *D*. Join *AD* and *CD* to complete the rhombus.

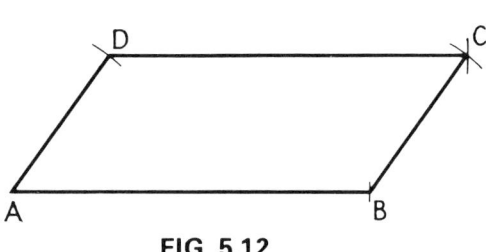

FIG. 5.12

FIG. 5.12 *To construct a parallelogram given the length of two sides and one angle, AB, BC, and ABC.*

Draw side *AB* and angle *ABC*. The parallelogram only differs from the rectangle by not having right angles, therefore the remainder of the construction is that of a rectangle. With centre *B* and radius side *BC*, describe an arc to cut *BC* at *C*. With centre *A* and radius side *BC*, describe an arc on which will be the point *D*, then with centre *C* and radius *AB* describe an arc to cut the one from centre *A* at *D*. Join *AD* and *CD* to complete the parallelogram.

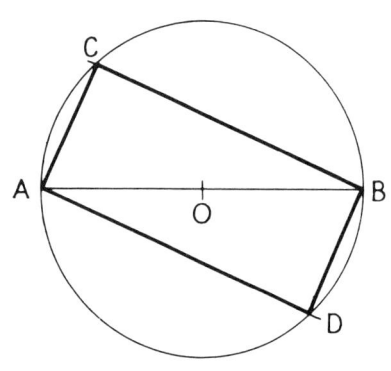

FIG. 5.13

FIG. 5.13 *To construct a rectangle given the length of the diagonal AB and one side AC.*

AB is the common side of two right-angled triangles, which together form the rectangle, therefore make *AB* the diameter of a circle and bisect *AB* and describe the circle. With centre *A* and radius side *AC*, describe an arc to cut the circumference at *C*. With centre *B* and radius side *AC*, describe an arc to cut the circumference at *D*. Join *AC*, *BC*, *BD*, and *AD* to complete the rectangle.

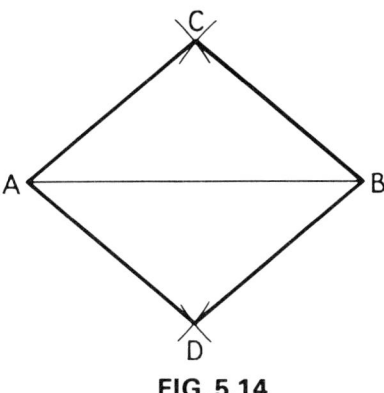

FIG. 5.14

FIG. 5.14 *To construct a rhombus given the length of a diagonal AB and the length of a side AC.*

This figure consists of two isosceles triangles with *AB* as the common base. Draw side *AB* and on it construct the two triangles, using *A* as centre and side *AC* as radius to describe arcs above and below *AB*. With *B* as centre and the same radius, describe arcs to cut the first arcs at *C* and *D*. Join *AC*, *AD*, *BC*, and *BD* to complete the rhombus.

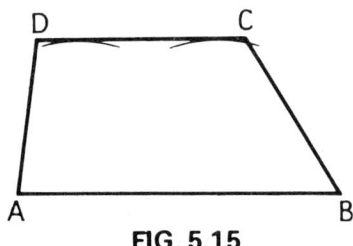

FIG. 5.15

FIG. 5.15 *To construct a trapezium given the perpendicular distance between the parallels, the sides AB and CD, and the angle ABC.*

Draw side *AB* and angle *ABC*. Construct a parallel to *AB*, the given perpendicular distance from *AB*, and produce it to give the exact position of *C*. With *C* as centre and radius side *CD*, step off point *D*. Join *AD* to complete the trapezium.

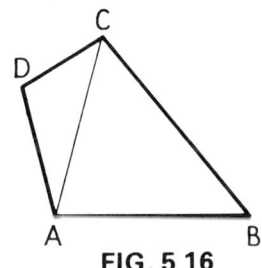

FIG. 5.16

FIG. 5.16 *To construct an irregular quadrilateral given AB, BC, AC, AD, and CD.*

This consists of two triangles *ACD* and *ABC*; the side *AC* is common. By the method of Fig. 5.1, draw side *AB* and on it construct triangle *ABC*. On *AC* construct the triangle *ACD* to complete the irregular quadrilateral.

EXERCISES

1. Construct the following triangles: (i) an equilateral of 60mm side, (ii) an isosceles of 60mm base and 90mm altitude, (iii) a scalene, $AB = 83$mm $BC = 48$mm, $AC = 60$mm, (iv) a right-angled of 60mm base and 90mm side.

2. Construct triangles given the following perimeters and proportions. (i) Perimeter 195mm, proportion 2:4:5. (ii) Perimeter 225mm, proportion 3:4:5. (iii) Perimeter 212mm, proportion 6:4:3.

3. Construct triangles with the following base angles and altitude. (i) Altitude 67mm, base angles 55° and 75°. (ii) Altitude 83mm, base angles 50° and 85°.

4. Construct equilateral triangles of altitude, (i) 70mm, (ii) 85mm.

5. Construct right-angled triangles, (i) one side 65mm, hypotenuse 110mm, (ii) one side 48mm hypotenuse 125mm.

6. Construct a triangle, altitude 70mm, vertical angle 50°, and base angle 55°.

7. Construct the following:

   (i) A square of 70mm side.

   (ii) A rectangle of 75mm and 110mm.

   (iii) A rhombus of 78mm side and 75° angle.

   (iv) A parallelogram of 75mm and 105mm sides with an included angle of 65°.

   (v) A rectangle of 120mm diagonal and one side 45mm.

   (vi) A rhombus of 60mm side and 90mm diagonal.

   (vii) A trapezium *ABCD*. $AB = 75$mm, $CD = 90$mm, $BC = 45$mm, $ABC = 85°$. *AB* and *CD* are parallel.

   (viii) An irregular quadrilateral *ABCD*. $AB = 52$mm, $BC = 60$mm, $AC = 75$mm, $CD = 37$mm, and $AD = 52$mm.

CHAPTER 6

CONSTRUCTION OF REGULAR POLYGONS

Regular Polygons are plane figures with equal sides and angles and named according to the number of sides. Two particular examples of their application are the standard hexagonal nut, used in engineering, and the mansard roof truss, used in building.

Polygons (without the word Regular) can be figures with unequal sides and angles, but in all examples the dimensions usually show whether a polygon is or is not a regular figure.

Common Polygons: Pentagon   5 sides.
Hexagon   6 sides.
Heptagon   7 sides.
Octagon   8 sides.
Nonagon   9 sides.
Decagon   10 sides.

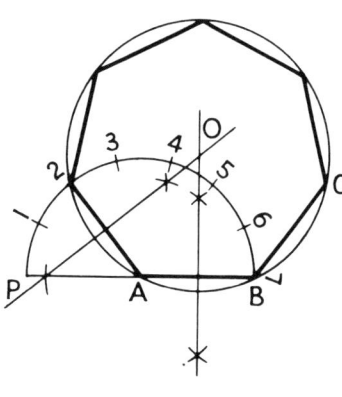

FIG. 6.1

FIG. 6.1 *To construct any regular polygon given the length of one side AB (Method 1).*

Draw side *AB* and, on *AB* produced, describe a semicircle of radius *AB*.

By a method of trial and error—stepping off chords with dividers—divide the circumference into the number of equal parts the polygon has sides, e.g. seven for a heptagon. Join the second point, 2, to *A* (always join the second point for any polygon), which gives the second side of the polygon, also *A*2 equals *AB* because they are radii of the same circle. Bisect the sides *A*2 and *AB* and produce the bisectors to meet at *O*, which is the centre of the circumscribing circle. With centre *O* and radius *OA* describe the circle. From *B* step off the length of side *AB* along the circumference and join these points.

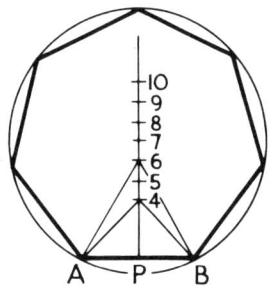

FIG. 6.2

FIG. 6.2 *To construct any regular polygon given the length of one side AB (Method 2).*

Draw side *AB*, bisect it to give the mid-point *P* and erect a perpendicular to *AB* at *P*. From *A* and *B* draw angles of 45° and 60° to intersect at 4 and 6; these give the centres of the circumscribing circles of a square and hexagon respectively, therefore bisect 4–6 to find the centre of a circumscribing circle of a pentagon. With the length 4–5 (or 5–6) step off points 7, 8, 9, 10, which—by the same reasoning—will be the centres of the circumscribing circles of a heptagon, octagon, nonagon, and decagon respectively. The construction of a heptagon is completed by describing a circle from centre 7, of radius 7*A*,,then from *B* stepping off the distance *AB* on the circumference.

Method 1 depends upon trial and error for the division of the semicircle compared with Method 2, which is entirely a geometrical construction.

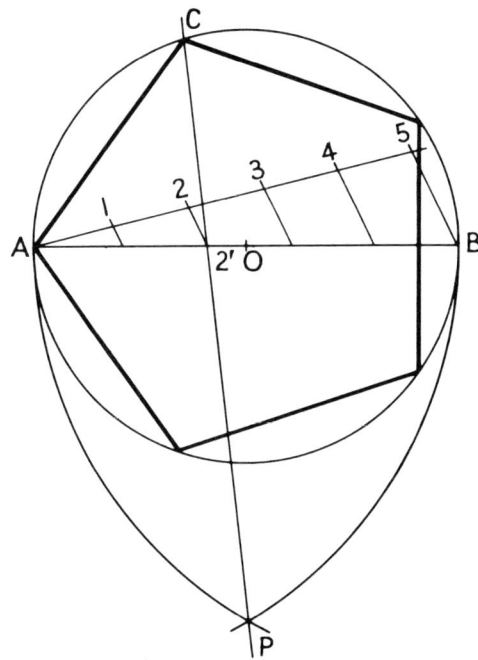

FIG. 6.3

FIG. 6.3 *To construct any regular polygon within a given circle of diameter AB.*

Draw the circle and its diameter *AB*. By the method of Chapter 4, divide the line *AB* into the number of equal parts the polygon has sides, in this example five, i.e. a pentagon. With *A* and *B* as centres describe two arcs, of radius *AB*, to intersect at *P*. Join point 2 on the diameter to *P* and produce the line to cut the circumference at *C* (the second point is used for any polygon). *AC* is one side of the pentagon, therefore from *A* step off the distance *AC* on the circumference and join the points to complete the polygon.

Special Cases

FIG. 6.4 *To construct a regular octagon within a given square ABCD.*

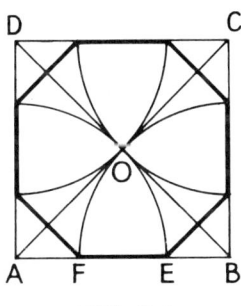

FIG. 6.4

Draw the square and its diagonals. With centres *A*, *B*, *C*, and *D* and radius half a diagonal (*AO*), describe four quadrants to cut the sides of the square and join these points to complete the octagon. By this method eight isosceles triangles, with vertices of 45°, are partly formed and their intersections form similar triangles each one-eighth of the octagon, e.g. triangle *AEO* cut by triangle *FBO* gives triangle *FEO*, one-eighth of octagon.

Given one side an octagon can be constructed with with a 45° set square.

FIG. 6.5 *To construct a regular hexagon given one side AB.*

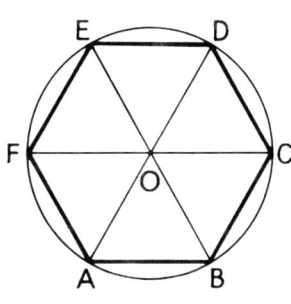

FIG. 6.5

Draw side *AB* and on it construct an equilateral triangle *ABO*. With *O* as centre and radius *OA* describe a circle. There are six equilateral triangles in a regular hexagon and dividing a circle into six equal parts and drawing the chords for each part also gives six equilateral triangles. Using the method of the division of a circle into equal parts (Fig.3.7), draw a parallel diameter, *CF*, to side *AB* and produce *AO* and *BO* with a 60° set square to *D* and *E*. Draw the chords *AF*, *BC*, *CD*, *ED*, and *EF* to complete the hexagon.

EXERCISES

1. On a given side of 37mm, construct the following regular polygons: (i) a pentagon, (ii) a hexagon, (iii) a heptagon, (iv) an octagon.

2. Within a circle of 90mm diameter, construct: (i) a pentagon, (ii) a hexagon, (iii) a heptagon, (iv) an octagon.

3. Within an 82mm square construct a regular octagon

4. Construct a regular hexagon of 35mm side.

5. Construct a regular hexagon of 82mm diagonal.

6. Construct a regular octagon of 98mm diagonal.

7. The altitude of the isosceles triangles, which together form a regular polygon, is 41mm; construct: (i) a pentagon, (ii) a heptagon, (iii) an octagon.

CHAPTER 7

REDUCTION AND ENLARGEMENT OF PLANE FIGURES

Draughtsmen do not merely draw but also design, and so must be able to calculate areas of figures and draw similar figures in fixed ratio to side or area.

The areas and reduction, or enlargement, of irregular figures are more easily found by geometrical rather than other mathematical methods. For instance, the shape of a figure can be quickly reduced to a rectangular strip of unit width and equal area, also the shape of a moulding at a mitre can be found by the enlargement method.

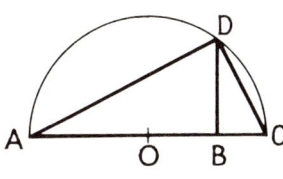

FIG. 7.1

FIG. 7.1 *To show the principle by which a mean proportional can be found.*

Figure 7.1 is a semicircle of centre O and diameter AC. D is any point on the circumference and ADC is a right-angled triangle (by the reasoning given for Fig. 5.7). BD is perpendicular from D to AC and forms two right-angled triangles ABD and BDC, therefore angle BDA + angle BAD is equal to $90°$, angle BDC + angle BCD is equal to $90°$. Since the angles BDA and BDC form the right angle ADC, then angle BDA is equal to angle BCD, angle BAD is equal to angle BDC, and the triangles ABD and DBC are similar. In two similar triangles (as in Chapter 4),

$$\frac{\text{side } AB}{\text{side } BD} = \frac{\text{side } BD}{\text{side } BC}$$

and, by multiplying, sides $AB.BC = $ side BD^2.

BD is the mean proportional of $AB.BC$, and by this method, if the lengths of the two sides of a rectangle are given, it is possible to draw an equal area square.

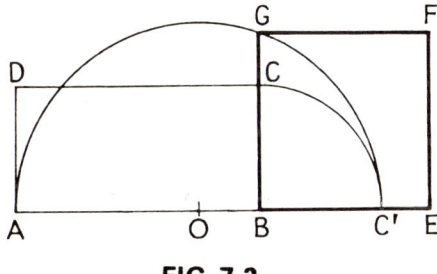

FIG. 7.2

FIG. 7.2 *To construct a square of equal area to a given rectangle ABCD.*

A construction of this kind depends upon finding the mean proportional to the two sides of a rectangle. To apply this principle, make the diameter of the semicircle $AB + BC$ (two sides of the rectangle), draw the semicircle, and produce BC to G on the circumference, then BG is the mean proportional and therefore the side of a square of equal area. Complete the square $BEFG$ (see p.13).

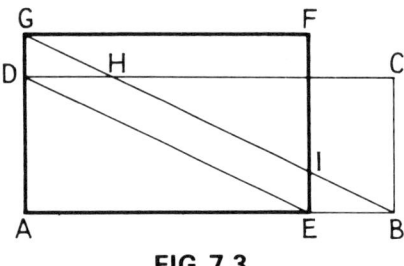

FIG. 7.3

Fig. 7.3 *To construct a rectangle of side AE and of equal area to a given rectangle ABCD.*

Draw the given rectangle by the method of Fig. 5.10, and produce the side *AD*. Make *AE* equal to the given side of the new rectangle and join *DE* to form the triangle *ADE*. From corner *B* draw a parallel line to *DE*, and let it cut *AD* produced at *G*. Erect a perpendicular, to *AB* at *E*, and from *G* draw a parallel line to *AB* to cut the perpendicular at *F*. Triangle *HCB* is equal to triangle *ADE*, because *DE* and *HB* are parallel, within the same rectangle and therefore equal; similarly triangle *GFI* is equal to triangles *ADE* and *HCB*. By this reasoning, *HB* is equal to *GI*, *GH* to *IB*, and triangle *DGH* to triangle *EIB* (both triangles are between the same parallels). Rectangle *AEFG* is equal in area to rectangle *ABCD*.

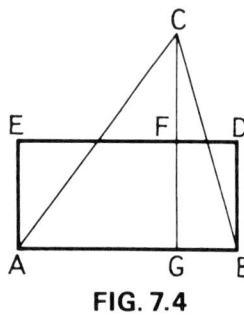

FIG. 7.4

Fig. 7.4 *To construct a rectangle equal to a given triangle ABC.*

Draw the triangle *ABC* and a perpendicular to *AB* from *C* (altitude *CG*). Because the area of a triangle is half the altitude times the base, i.e. $\frac{1}{2}CG \times AB$, bisect *CG* at *F*, and through *F* draw a parallel to *AB*, and complete the rectangle *ABDE*.

Fig. 7.5 and 7.6 *To construct a triangle of equal area to any given heptagon. In this example the heptagon ABCDEFG.*

Fig. 7.5 Stage 1. *Reduction to an equal area pentagon.*

Draw the heptagon *ABCDEFG*. Form the triangles *DBC* and *FAG* by joining *BD* and *AF*. Draw *CI* parallel to *BD* and *GH* parallel to *AF*, both to meet *AB* produced. Since triangles on the same base and between the same parallels are equal in area, join *DI* to form triangle *DBI*, equal in area to *DBC*, and *FH* to form triangle *FAH*, equal in area to *FAG*. The pentagon, *IDEFH*, is equal in area to the heptagon.

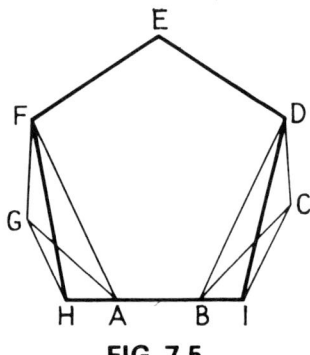

FIG. 7.5

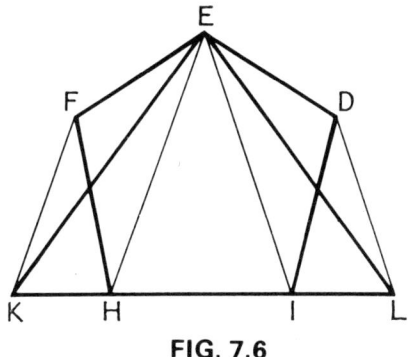

FIG. 7.6

FIG. 7.6 Stage 2. *Reduction to an equal area triangle.*

By the same reasoning as in stage 1, triangles of equal area can be constructed in the pentagon *IDEFH*. Join *EH* and *EI*, draw *FK* parallel to *EH* and *DL* parallel to *EI*, both to meet *IH* produced. Join *EK* and *EL* to complete the triangle *EKL*. Triangle *EHF* is equal to *EHK* and triangle *EID* is equal to *EIL*, therefore triangle *ELK* is equal in area to the pentagon and heptagon.

This is a general method for any polygon.

FIG. 7.7 *To construct a similar figure of a given ratio of side to a given figure ABCDE* (i.e. an enlarged figure of ratio 2:1)

This method is based on the division of lines into a given ratio (Fig. 4.2). Draw the given figure *ABCDE* From a convenient point *P* draw radial lines through *A, B, C, D, and E*. Produce *AE* to *H* in the given ratio (directly in this example of 2:1, but by means of a scale for more difficult ratios). Through *H* draw a parallel to *AP* produced to intersect with *EP* produced at *E'*. Draw *A'E'* parallel to *AE*, *A'B'* parallel to *AB*, *B'C'* parallel to *BC*, *C'D'* parallel to *CD*, and *D'E'* parallel to *DE*. In this construction it will be noticed that two similar triangles are drawn to give one enlarged side.

Reduction depends upon the same principle, but differs by the point *H* being between *A* and *E*, and the reduced figure being nearer to point *P* than the given figure.

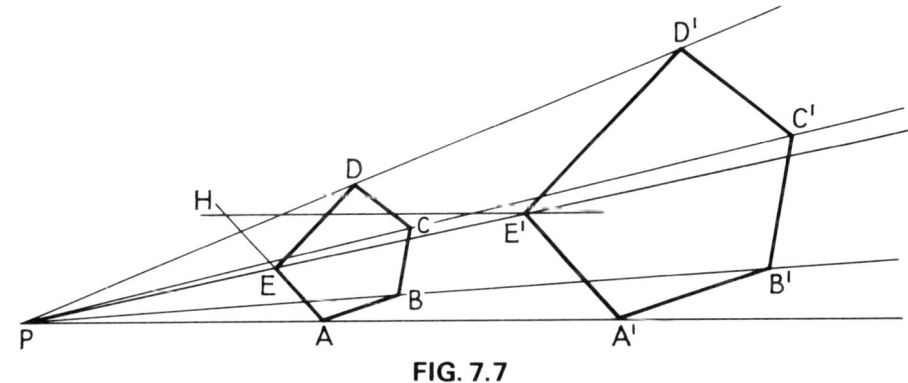

Enlargement and Reduction

FIG. 7.7

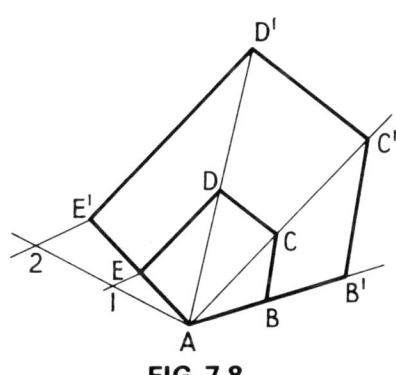

FIG. 7.8

FIG. 7.8 *To construct a similar figure of a given ratio of side to a given figure ABCDE.*

This is an alternative method to that used in Fig. 7.7. The principle of construction again depends upon the formation of similar triangles by lines all drawn from one point and passing through the corners of the figure. Draw the given figure *ABCDE* and a proportional scale (as in Chapter 4) on the side *AE* to enlarge *AE*, in this example in a ratio of 2:1. Draw radial lines from *A* through *D* and *C*. Produce *AE* to the new length *AE'* and, from *E'*, draw a parallel to *ED* to cut the radial *AD* at *D'*. Similarly draw *D'C'* and *C'B'* parallel to *DC* and *CB* respectively. Produce *AB* to *B'* to complete the enlarged figure.

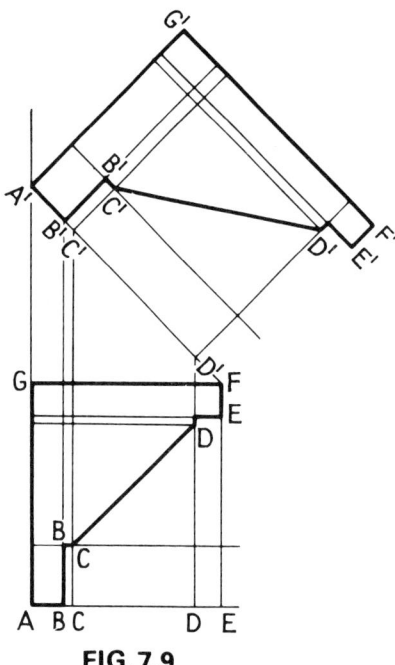

FIG. 7.9

Fig. 7.9 *To construct a figure larger in one direct-tion than a given figure ABCDEFG, given F'G' greater than FG and A'G' equal to AG.*

Draw the given figure *ABCDEFG* and produce *AG*. With *F* as centre and radius *F'G'* describe an arc to cut *AG* produced at *A'*. Join *A'F* and draw parallels to *AG* through *B, C,* and *D* to cut *A'F* at *B', C',* and *D'*. From *A', B', C',* and *D'* draw perpendiculars to *A'F* and transfer the lengths *AG* to *A'G', BB* to *B'B', DD* to *D'D',* and *EE* to *FE'*. Draw parallel lines to *A'F* through *C', E',* and *G'* and complete the enlarged figure.

This enlargement is often used in drawing the shapes of metal or wood brackets, e.g. stair brackets. This kind of enlargement can be produced, in Solid Geometry, by means of a section plane.

Ratio of Area

The fact that $AB^2 + AC^2 = BC^2$ in any right-angled triangle is applied not only to squares but to similar figures and is used, in addition to the mean proportional, in finding the ratio of areas. In Fig. 7.10, on the sides of the right-angled triangle, are three similar figures, the areas having the same relationship, as the squares on the sides, to the triangle. Briefly, area of figure on *AB* + area of figure on *AC* = area of figure on *BC*, i.e. $AB^2 + AC^2 = BC^2$.

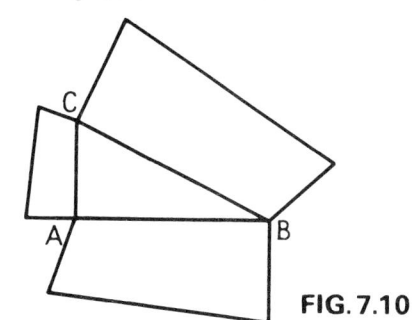

FIG. 7.10

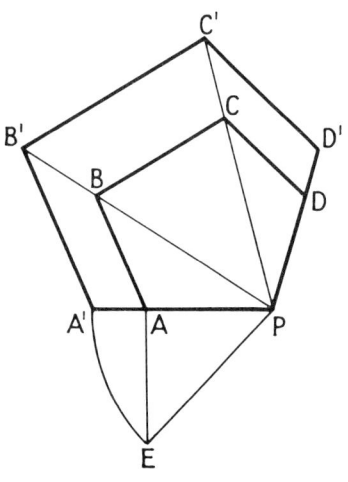

FIG. 7.11

Fig. 7.11 *To construct a similar pentagon of twice the area of the given pentagon ABCDP.*

Draw the pentagon *ABCDP* and construct a right-angled triangle on *AP*, making *AE = AP*. The enlarged side of *AP* will be equal to the side of *EP* of the right-angled triangle; therefore produce *PA* and with centre *P*, radius *EP*, describe an arc to give point *A'* (the line *EP* has been swung round or *rabatted* to *A'P*). From *P* draw radials through *B* and *C* and produce *PD*. From *A'* draw *A'B'* parallel to *AB*, *B'C'* parallel to *BC*, and *C'D'* parallel to *CD*, completing the enlarged figure by the principle of similar triangles.

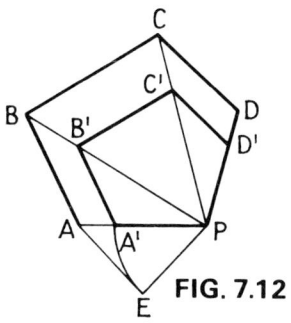

FIG. 7.12 *To construct a similar pentagon of half the area to a given pentagon ABCDP.*

Draw the given pentagon *ABCDP*. With *AP* as the hypotenuse, construct a right-angled triangle making *AE = EP* (*EP* will be the length of the reduced side *AP*). Rabat *EP* to *A'P*, then draw radials from *P* to *B* and *C*. From *A'* draw *A'B'* parallel to *AB*, *B'C'* parallel to *BC*, and *C'D'* parallel to *CD*.

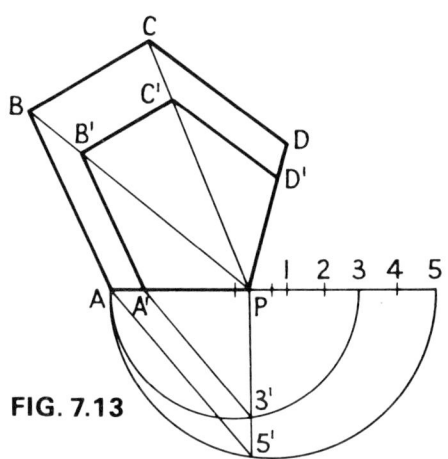

FIG. 7.13 *To construct a similar pentagon, of ratio 3:5 in area, to a given pentagon ABCDP.*

This construction is more complicated in the drawing of a right-angled triangle to give squares of the correct ratio, but since a mean proportional is the square of two measurements then it can be combined with the right-angled triangle to produce the required figure. Draw the pentagon *ABCDP* and produce *AP* . From *P* and on *AP* produced step off five equal spaces. Describe semicircles on *A*3 and *A*5. Draw *P*3′5′ perpendicular to *AP*; therefore *P*5′ and *P*3′ are mean proportionals to *AP* and *P*5, and *AP* and *P*3 respectively. Join *A* to 5′, and draw *A*′3′ parallel to *A*5′ to complete two right-angled triangles. Because the triangles are similar then *A'P* is the reduced side of *AP* (the correct ratio is found by drawing the mean proportional). By parallels and radials draw *A'B'*, *B'C'*, and *C'D'*.

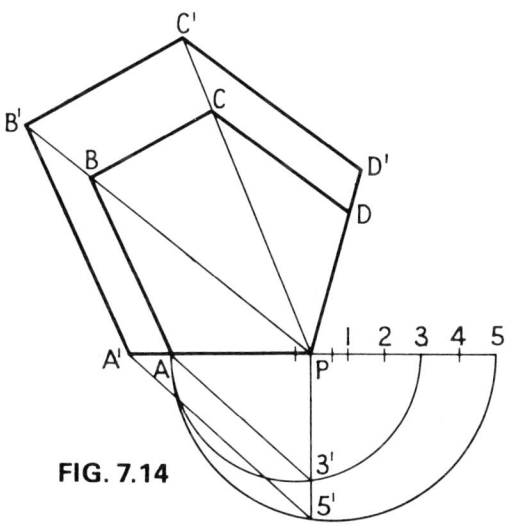

FIG. 7.14 *To construct a similar pentagon, of ratio 5:3 in area, to a given pentagon ABCDP.*

The principle is the same as that of Fig. 7.11, therefore draw *ABCDP* and produce *AP*. From *P* step off five equal spaces on *AP* produced and describe semicircles on *A*3 and *A*5. Draw *P*3′5′ perpendicular to *AP*. Join *A* to 3′ and *A*′5′ parallel to *A*3′. Complete the pentagon through *A'B'C'D'*.

These are general methods and applicable to all polygons.

EXERCISES

1. Construct squares equal in area to the following rectangles: (i) 90mm × 37mm, (ii) 75mm × 22mm, (iii) 82mm × 26mm (iv) 64mm × 34mm.

2. Construct rectangles equal in area to other rectangles, given one side of the required rectangle: (i) side 37mm, rectangle 49mm × 56mm, (ii) side 49mm, rectangle 60mm × 42mm, (iii) side 52mm, rectangle 75mm × 37mm.

3. Convert the following triangles into rectangles of equal area: (i) equilateral of 45mm side, (ii) isosceles of 52mm base, 75mm altitude, (iii) right-angled of hypotenuse 75mm, one side 52mm.

4. Reduce a pentagon, hexagon, heptagon, and octagon, all of 37mm side, each to an equal area triangle.

5. A pentagon $ABCDE$ has the following dimensions: $AB = 60$mm, $BC = 45$mm, $CD = 52$mm, $DE = 67$mm, $AE = 37$mm, and $AC = AD = 90$mm. Construct the polygon and reduce it to an equal area square.

6. Construct a similar pentagon, of ratio of side 5:3, to a regular pentagon of 30mm side.

7. Construct a similar pentagon, of ratio of side 2:5, to a regular pentagon of 60mm side.

8. Draw the irregular pentagon of Exercise 5.
(i) Enlarge in ratio of side 7:4.
(ii) Reduce in ratio of side 1:2.

9. The dimensions of Fig. 7.9 are $AG = 96$mm; $GF = 80$mm; $AB = 16$mm; $BC = 4$mm, $DE = 12$mm; $FE = 16$mm; $DD = 80$mm; $BB = 28$mm. Draw the figure and enlarge it in one direction:
(i) Make $FG = 120$mm, AG to remain 96mm.
(ii) Make $AG = 135$mm, FG to remain 80mm.

10. Construct similar polygons of twice the area to the following, all of 30mm side: regular pentagon, regular hexagon, regular heptagon, and a regular octagon.

11. Construct similar polygons of half the area to a regular pentagon, a regular hexagon, a regular heptagon, and a regular octagon, all of 45mm side.

12. Construct a pentagon $ABCDE$, where $AB = 45$mm, $BC = CD = 34$mm, $DE = 42$mm, $AE = 37$mm, $AD = 60$mm, and $BD = 52$mm. Draw a similar figure: (i) twice the area; (ii) half the area; (iii) ratio of area 5:3; (iv) ratio of area 7:4; (v) ratio of area 3:4; (vi) ratio of area 4:7.

CHAPTER 8

TANGENTS AND CIRCLES OF A TRIANGLE

Any straight line which touches a circle at a point on the circumference, and does not cut it, though it be produced, is a tangent, and if this line is not continued beyond the point, then the tangent and the circumference form an unbroken line. In drawing shapes consisting of curved and straight lines, e.g. bearings, small castings, and brick arches, the constructions in this chapter are essential.

FIG. 8.1

FIG. 8.1 *To construct a tangent from a given point P to a given circle, centre O, radius OA.*

The tangent will pass through one point on the circumference, but it is also perpendicular to a radius—the perpendicular radius to a tangent is called a normal—therefore the tangent and normal form two sides of a right-angled triangle whose hypotenuse is the line joining P and the centre of the circle O. Mark the position of P and describe the circle OA. Using the principle of Fig. 5.7, describe a semicircle on OP and join T (the point of intersection of the two curves) to O and P to form a right-angled triangle in the semicircle; then OT is the normal and PT produced is the required tangent.

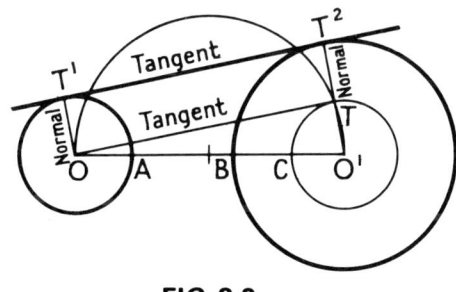

FIG. 8.2

FIG. 8.2 *To construct an external tangent to two given circles, centres O and O', radii AO and BO' respectively.*

Draw the given circles AO and BO' and a third circle with centre O' and radius CO', i.e. making CO' the difference between AO and BO'. Construct the tangent OT from centre O to circle CO', as in Fig. 8.1. Produce the normal $O'T$ to T^2, draw a parallel normal from O to T^1, join and produce $T^1 T^2$ which is the required tangent touching both circles and also equidistant from the tangent OT, because BC is equal to AO.

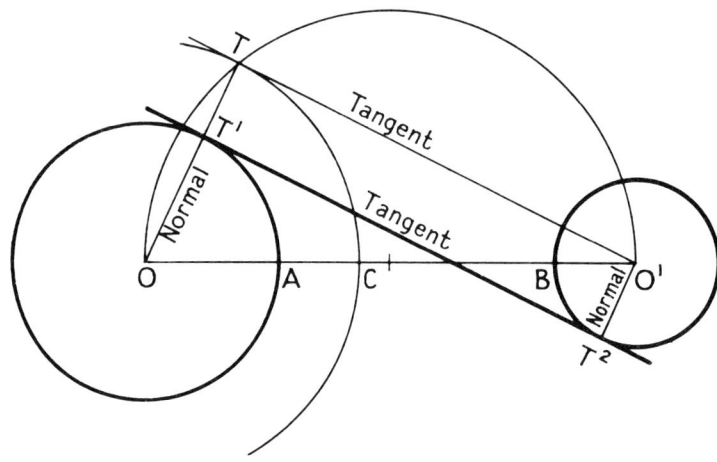

FIG. 8.3

FIG. 8.3 *To construct an internal tangent to two given circles, centres O and O', radii AO and BO'
respectively.*

The difference between the construction of an internal tangent and an external one is in the radius of
the circle of the first tangent OT; this is the sum of the radii of the given circles, $AO + BO'$, instead of the
difference. Draw the given circles AO and BO' and a third circle with centre O and radius CO, i.e. $AO+BO'$.
Construct the tangent $O'T$ from centre O' to circle CO and draw a parallel normal to OT from O' to give
point T^2. Join $T^1 T^2$ which is the required tangent and equidistant from the tangent $O'T$.

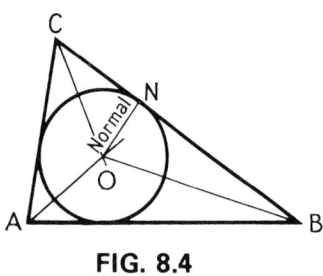

FIG. 8.4

FIG. 8.4 *To draw the inscribed circle of a given
triangle ABC.*

The sides of the triangle will be tangents to the
circle and since these sides (tangents) meet at the
three points ABC, then the centre of the inscribed
circle must lie on the bisectors of these three angles.
Draw the triangle ABC, and bisect two internal
angles to give point O. Draw the normal ON and,
with O as centre and radius ON, describe the circle.

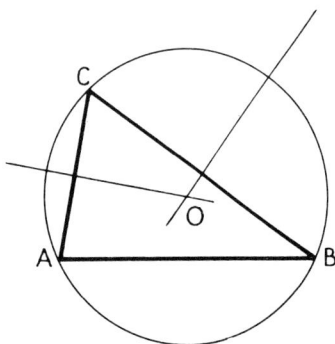

FIG. 8.5

FIG. 8.5 *To draw the circumscribed circle of a given
triangle ABC.*

Draw the triangle ABC, the sides of which will
be the chords of the circumscribing circle; therefore
bisect any two sides and produce the bisectors to
intersect at O. With O as centre and radius AO,
describe the circle.

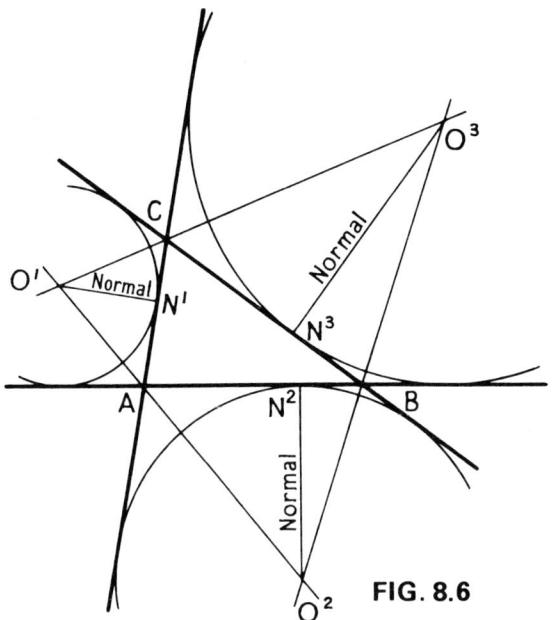

FIG. 8.6

FIG. 8.6 *To draw the escribed circles of a given triangle ABC.*

Draw the given triangle *ABC* and produce the sides in both directions. The sides and sides produced are tangents to the escribed circles and therefore the centres will be on the bisectors of the angles formed by the sides and sides produced, using the same reasoning as in the case of the inscribed circles. Bisect the external angles of the triangle and produce these bisectors to meet at O^1, O^2, and O^3. Draw the normals N^1O^1, N^2O^2, and N^3O^3, each perpendicular to its tangent (i.e. side). With centres O^1, O^2, and O^3 and radii N^1O^1, N^2, O^2, and N^3O^3 respectively, describe the three circles.

EXERCISES

1. From a point *P* draw a tangent to a circle:
(i) circle 45mm radius, *P* 75mm from the centre.
(ii) circle 52mm radius, *P* 80mm from the centre.

2. Two circles are of 50mm and 30mm radii respectively and their centres are 120mm apart:
(i) Draw the external tangents to the circles.
(ii) Draw the internal tangents to the circles.

3. Repeat Exercise 2 for circles 56mm and 18mm radii, centres 123mm apart.

4. Draw the inscribed, escribed, and circumscribed circles of a triangle *ABC*, where *AB* = 60mm *BC* = 56mm, and *AC* = 45mm.

5. Within a regular pentagon of 45mm side, describe five equal circles each touching each other and each touching one side of the pentagon.

6. About a circle of 60mm diameter, construct an equilateral triangle, and within the circle describe three equal circles each touching each other and one point on the circumference.

CHAPTER 9

CIRCLES AND STRAIGHT LINES

Once the properties of the tangent and normal are known it is possible to apply them practically, e.g. in drawing a train of gear wheels, the profile of a wooden moulding or a metal casting in which a circle is combined with a straight line. The following examples give the essentials for drawings of this kind.

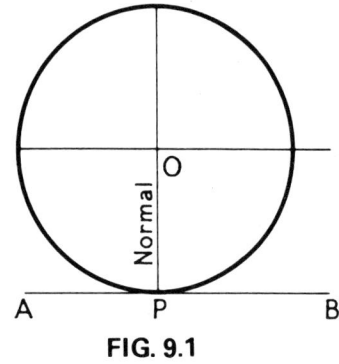

FIG. 9.1

FIG. 9.1 *To describe a circle of radius OP to pass through a point P on the straight line AB.*

Draw the tangent line *AB* and mark *P*. Draw the normal *OP* equal to the radius and with centre *O*, radius *OP*, describe the circle.

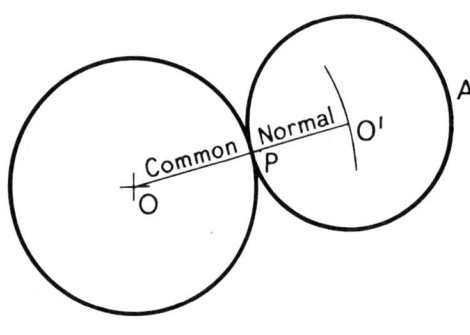

FIG. 9.2

FIG. 9.2 *To describe a circle of radius AO′ to touch a given circle, radius OP, and pass through the point P.*

Describe the circle of radius *OP* and mark *P*. The point of contact will be on a common tangent to both circles, therefore join *OP* and produce to form the common normal. With *O* as centre and radius *OP* + *AO′*, describe an arc to cut the common normal at *O′*, the centre of the required circle. With radius *AO′* describe the circle to pass through *P* the point on the common normal and circumference of the circle *OP*.

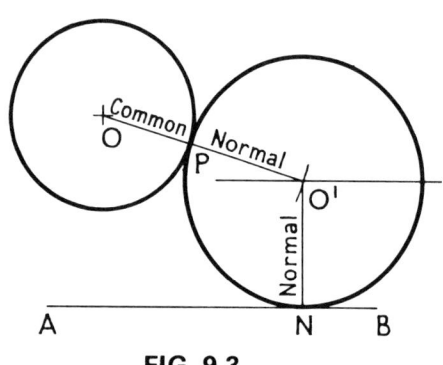

FIG. 9.3

FIG. 9.3 *To describe a circle of radius NO′ to touch a given straight line AB, and a given circle, radius OP.*

Describe a given circle of radius *OP* and draw *AB*. The centre of the circle will be the intersection of an arc (centre *O* and radius the given circle + *NO′*) with a parallel line a distance *NO′* from the circle's tangent *AB*. Draw the parallel line a perpendicular distance *NO′* from *AB* and describe an arc with centre *O* and radius *OP* + *NO′* to give centre *O′*. Draw the normals and describe the circle of radius *NO′* to pass through the common normal at *P* and meet *AB* (as a tangent) at *N*.

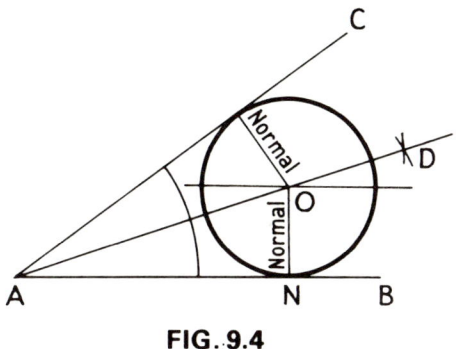

FIG. 9.4

FIG. 9.4 *To construct a circle of radius NO to touch two converging lines AB and AC.*

The lines will be tangents to the circle and the construction is similar to the inscribed circle (Fig 8.4). Draw *AB*, *AC* and bisect angle *BAC*. The centre will be the intersection of the bisector *AD* and a parallel line a perpendicular distance *NO* from one tangent. Construct the parallel to *AB* to give centre *O*, draw the normals and describe the circle of radius *NO*.

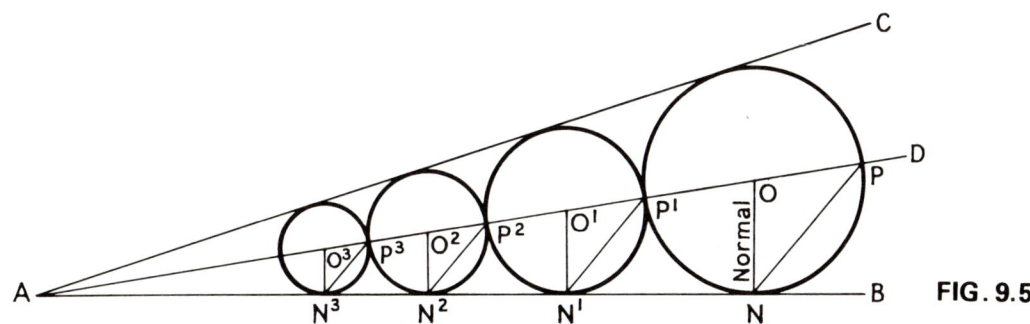

FIG. 9.5

FIG. 9.5 *To draw a series of circles, each touching the preceding circle and two converging lines, AB and AC.*

Draw *AB* and *AC*. As in Fig. 9.4 bisect the angle *BAC* to find the centre line (which is also the common normal) of all the possible circles. Draw a normal from any convenient point (*N* on *AB*) to give *O* on *AD*. Describe a circle with centre *O* and radius *NO*. Join *NP* to complete the triangle *ONP*. If a similar triangle is drawn at the opposite end of the diameter of the first circle, at *P¹* on the centre line *AD*, then *O¹* will be the centre of a circle which will touch both lines and the first circle. Draw *N¹P¹* parallel to *NP* and the normal *N¹O¹* to complete the similar triangle *O¹N¹P¹*. With centre *O¹*, radius *N¹O¹*, describe the second circle. The method is the same for additional circles.

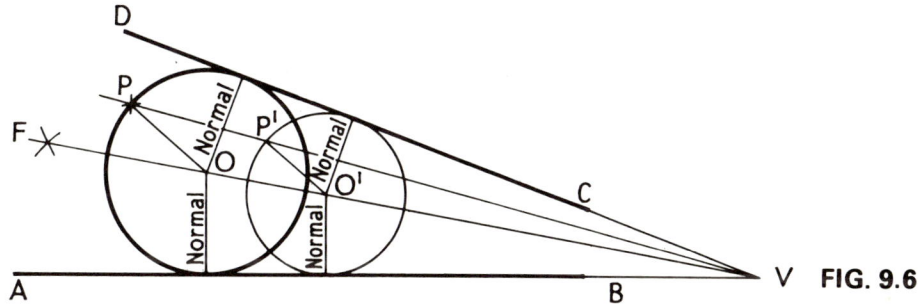

V   FIG. 9.6

FIG. 9.6 *To describe a circle to pass through a point P and touch two converging straight lines AB and CD. P is between the lines AB and CD.*

Draw the converging lines *AB* and *CD*, produce them to intersect at *V* and mark the position of *P*. The construction partly depends upon the method of describing a circle to touch two converging lines, Fig. 9.4. Find the bisector *VF* of the angle *AVD* and, at any convenient point *O'*, draw the normals and describe a circle to touch *AB* and *CD*. Join *P* to *V* to give the point *P'* on the circle of centre *O'*. Since a second circle must be drawn which will pass through *P* and also touch *AB* and *CD*, then a similar method to that used in Fig. 9.5 can be applied. Construct two similar triangles by joining *O'P'* and drawing a straight line through *P*, parallel to *O'P'*, to give the point *O*. With *O* as centre, radius *OP*, describe the required circle.

Similarly, a second circle to pass through *P* and touch *AB* and *CD* could have been described with its centre to the left of *P*.

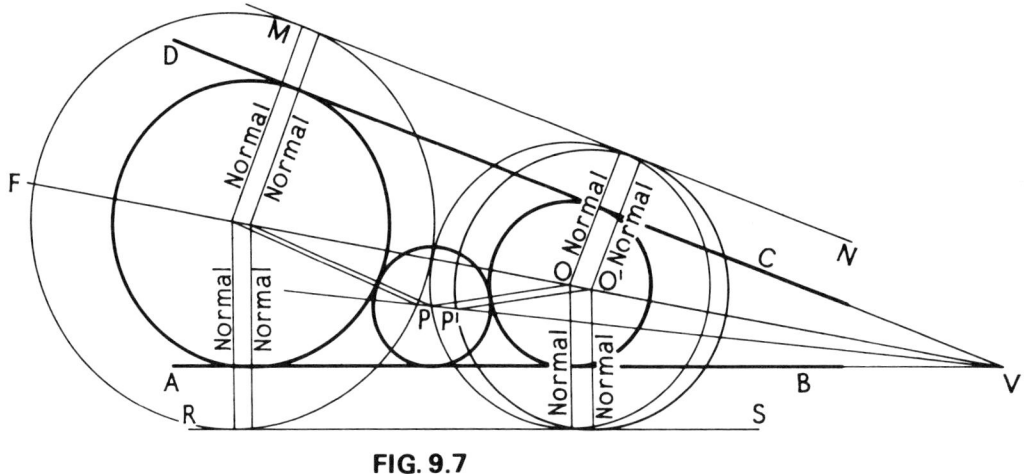

FIG. 9.7

Fɪɢ. 9.7 *To draw two circles to touch two converging lines, AB and CD, and a given circle centre P. The given circle lies between AB and CD.*

Draw *AB*, *CD*, and the given circle of centre *P*. Draw the parallel lines *RS* and *MN*, a perpendicular distance equal to the radius of the given circle of centre *P*, from *AB* and *CD* respectively. Produce *AB* and *CD* to intersect at *V* and draw the bisector, *VF*, of the angle *AVD*. Since the distance between *MN* and *CD*, *RS* and *AB*, is equal to the radius of the given circle, then a circle described to touch *MN* and *RS* and pass through centre *P* will be concentric with a circle which will touch *AB*, *CD* and the given circle of centre *P*. As in Fig. 9.6, at a convenient point *O'*, draw the normals and describe a circle to touch *RS* and *MN*. Join *PV* to find *P'*. Join *O'P'* and draw a parallel through *P* to find *O*, the centre of the circle which will pass through *P* and touch *RS* and *MN*. Draw the normals from *O* and describe the circle to pass through *P* and touch *RS* and *MN*. With the same radius, minus that of the given circle, and centre *O*, describe the first of the required circles to touch *AB*, *CD* and the circle of centre *P*.

*To draw the second of the required cir*cles, repeat the same construction to the left of the given circle, as shown in the figure.

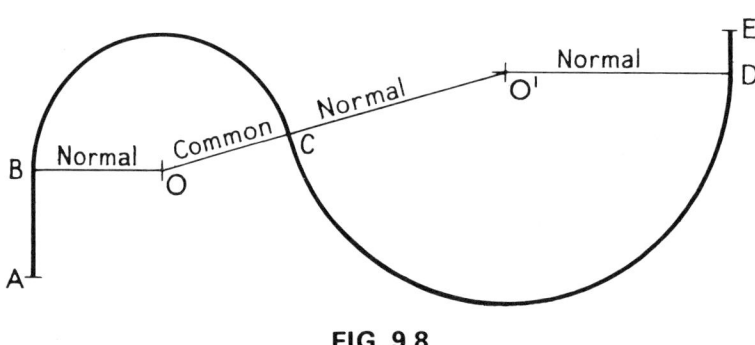

FIG. 9.8

Fɪɢ. 9.8 *To draw a continuous line consisting of a straight line AB, an arc from centre O and radius BO, a second arc from centre O' and radius CO', and a straight line DE parallel to AB (angle BOO' is 165°).*

Draw the straight line *AB* and the normal *OB* perpendicular to *AB* (because *AB* is the tangent to the arc of radius *OB*). From *O* draw the common normal *OO'*, making *OO'* equal to the radii *OB + O'C* and angle *BOO'* equal to 165°. From *O'* draw the normal *O'D*, parallel to *OB* and equal to the radius *O'C*. With centre *O* describe the arc of radius *OB* from *B* to *C* on the common normal. With centre *O'* describe the second arc of radius *O'C* from *C* on the common normal to *D*. Draw *ED* parallel to *AB*.

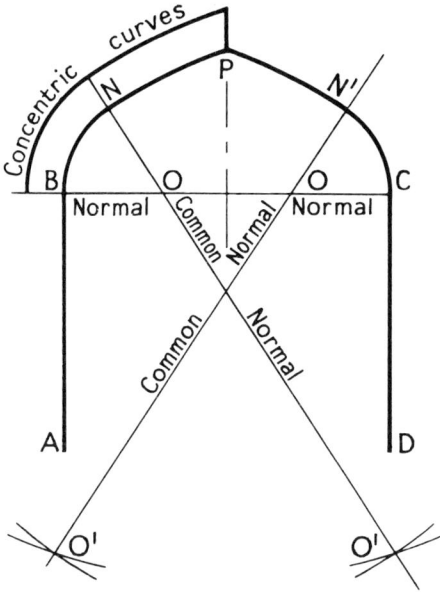

FIG. 9.9

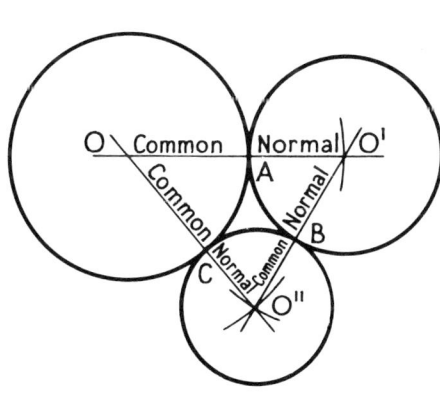

FIG. 9.10

FIG. 9.9. *To draw two equal and continuous lines, AP and DP, each consisting of one straight line and two arcs, P being a given vertical distance above A and D. The straight lines AB and CD are vertical, equal, and of distance AD apart. The arcs are of radii OB and O'P.*

Draw *AB* and *CD*, both tangents to the arcs of radius *OB*, and draw the normal *BC*. Since the lines *AP* and *DP* are equal, then *P* will be on a centre line between *AB* and *CD* and the given vertical distance above *BC*. Draw the centre line and mark *P*. Mark the centres *O*, radius *OB*, from *B* and *C*. Since the two curves will join on a common normal, this normal will pass through *O* and *O'* for each of the lines, therefore with *O* as centre, describe arcs of radius *O'P–OB*. With *P* as centre describe an arc of radius *O'P* to intersect the arcs from *O* and to give the centres *O'*. With centres *O* and radius *OB*, describe arcs from *B* to *N* and *C* to *N'* (*N* and *N'* are the points of contact of the arcs, common tangents, and on the common normals). With centres *O'* and radius *O'P*, describe arcs from *N* to *P* and *N'* to *P* to complete the two lines. The completed lines could form an arch, and concentric curves could be drawn, as indicated, to show the thickness.

FIG. 9.10 *To draw three circles of radii OA, O'B, and O"C to touch one another externally.*

With *O* as centre and radius *OA* describe the first circle. With *O* as centre and radius *OA + O'B* draw an arc and from *O'*, any point on the arc, draw a common normal to *O*. With *O'* as centre and radius *O'B* describe the second circle to pass through *A*. With *O* as centre and radius *OA + O"C* describe an arc and, with *O'* as centre and *O'B + O"C* as radius, describe a second arc to intersect the first at *O"*. Draw the common normals *O"O* and *O'O"*. With *O"* as centre and radius *O"C* describe a circle to pass through *C* and *B*.

EXERCISES

1. Draw a circle to touch a straight line *AB* 90mm in length at a point *P*: (i) radius 23mm, *P* 30mm from *A* on the straight line *AB*; (ii) radius 37mm, *P* 52mm from *A* on the straight line *AB*.

2. A circle is of 37mm radius and has a point *P* on its circumference. Draw a circle to touch this circle at point *P*, (i) 27mm radius, (ii) 45mm radius.

3. The centre of a 45mm diameter circle is 60mm vertically above a horizontal straight line. Draw a circle to touch this circle and the straight line, (i) 45mm radius, (ii) 27mm radius.

4. Two lines converge to a point; draw a circle to touch both lines, (i) 52mm diameter, angle of 30°, (ii) 68mm diameter, angle of 45°.

5. Two lines converge to a point. Draw three circles to touch both these lines, with the middle

circle touching the other two:

(i) Angle 20°, first circle 60mm diameter.

(ii) Angle 35°, first circle 45mm radius.

6. Two straight lines *AB* and *CD* converge and when produced intersect at a point *V*. The angle *AVD* is 25°.

(i) Draw a circle to touch both lines and pass through a point *P*, which is a perpendicular distance of 15mm from a point on *AB* 120mm from *V* and is between *AB* and *CD*.

(ii) Draw a circle to touch both lines and a given circle of centre *P* and radius 15mm.

(iii) Draw two circles to touch both lines and a given circle of radius 18mm, centre 18mm from a point on *AB* 145mm from *V*.

7. A continuous line consists of a 22mm vertical straight line, an arc of 27mm radius, a second arc of 34mm radius, and a 15mm vertical straight line. The common normal to the two curves makes an angle of 150° with the normal of the first curve to the 22mm straight line. Draw the line.

8. Draw Fig. 9.9 to the following dimensions: *AB* = *CD* = 45mm; *BC* = 165mm; *P* above *BC* = 60mm; *OB* = 37mm; and *O'P* = 202mm. Draw concentric curves to *BP* and *CP* of radii 67mm and 232mm. Show clearly your construction.

CHAPTER 10

ELLIPSE AND LOCI

Elliptical shapes are used in engineering and building, e.g. semi-elliptical arches (for a bridge, or a small building), and, under certain conditions, as the path of a point on a moving crankshaft. An ellipse is defined as the locus (or path) of a point moving under fixed conditions, and to construct it the characteristics of the curve must be known.

(i) The curve is symmetrical about two diameters or axes, the major and the minor, which bisect and are perpendicular to each other.

(ii) Two fixed points on the major axis, the foci, are equidistant from the minor axis, and the distance from either end of the minor axis to either focus equals half the major axis.

(iii) The sum of the distances from the two foci to any point on the ellipse is equal to the length of the major axis.

To construct an ellipse given the major and minor axes AB and CD.

FIG. 10.1 Method 1. *Drawing pins and thread.*

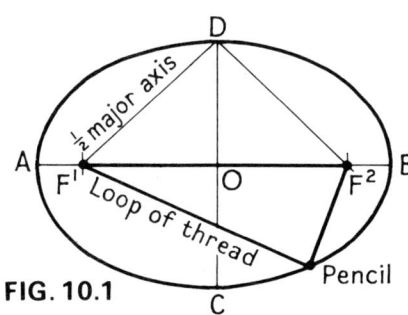

FIG. 10.1

This method is based on the definition of an ellipse; the pencil is the moving point, and the loop of thread, moving about two drawing pins placed at the foci, provides the fixed conditions. Draw the major and minor axes, AB and CD. On the major axis mark the foci, F^1 and F^2, by making C or D a centre and with a radius of half the major axis. Place the drawing pins in F^1 and F^2 and tie a piece of thread to form a loop from F^2 to A, or F^1 to B (the length of the thread, to the point where it is tied, is equal to $2 \times AF^2$ or $2 \times BF^1$). Place the pencil point in the loop at A (or B) and, keeping the loop taut, describe the curve. In describing the ellipse, the sum of the distances from the foci to the pencil is always the same.

FIG. 10.2 Method 2. *Rectangle.*

This method is very important because it can be applied to either a rectangle or a parallelogram. When applied to the latter the centre lines (which do not intersect to form right angles) become diameters. In Method 1 the stretching or varying tightness of the thread can cause irregularities in the ellipse, but in this method points are plotted geometrically and then a smooth freehand curve is drawn through them. Draw the major and minor axes, AB and CD, and a rectangle about their ends. Divide AOB into eight equal parts (or any convenient number) and divide EH and FG into the same number of equal parts, i.e. eight. From C draw lines through 1, 2, 3, 0, 3, 2, 1 on the major axis and repeat from D. From C draw lines to $1', 2', 3'$ on AE and BF. From D draw lines to $1', 2', 3'$ on AH and BG. Draw a smooth freehand curve through the points of intersection from D through AB with lines from AH and BG, and lines from C through AB with lines from AE and BF. Only half the construction is shown.

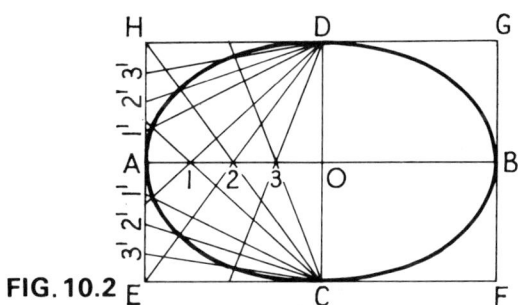

FIG. 10.2

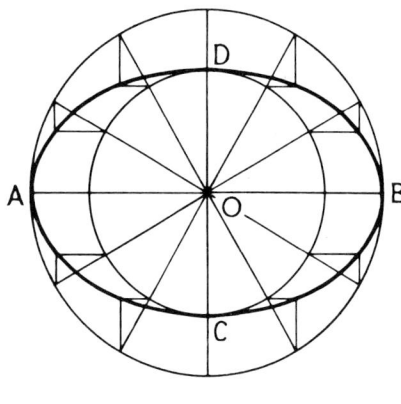

FIG. 10.3

Fɪɢ. 10.3 Method 3. *Circle.*

This is an alternative to Method 2. Draw the major and minor axes, *AB* and *CD*. With *O* as centre and radii *AO* and *CO*, describe two circles. Divide the circles into a number of equal parts, e.g. twelve, and from the divisions on the circumference of *AO* erect or drop perpendiculars to the major axis. From the divisions on the circumference of *CO* draw perpendiculars to the minor axis to intersect the perpendiculars on *AO*. Draw a smooth curve through the points of intersection.

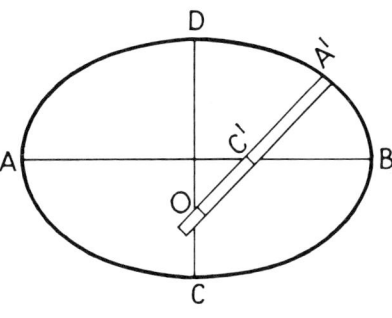

FIG. 10.4

Fɪɢ. 10.4 Method 3. *Trammel*

This is a locus method in which a strip of card moves along the major and minor axes, such that the distances from the end of the strip to the major and minor axes are always the same. Draw the major and minor axes, *AB* and *CD*. On a narrow strip of card mark *A'O* equal to half the major axis and *A'C'* equal to half the minor axis. With *C'* on the major axis for all positions and *O* on the minor axis for all positions, trace the path or locus of the point *A'* by moving the strip under these fixed conditions.

Normal and Tangent

The tangent to an ellipse has the same properties as that of a tangent to a circle, but the normal is different in that its position depends upon two points (the foci) instead of one, and it is the bi-

sector of the angle formed by the point through which the tangent passes and the foci. In Machine Drawing, the tangent and normal are necessary to draw shapes consisting of curves and straight lines and equidistant curves, e.g. a semi-elliptical stone or steel arch.

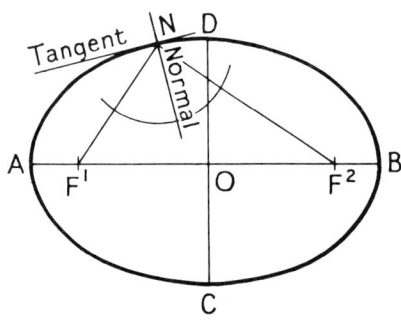

FIG. 10.5

Fɪɢ. 10.5 *To draw the normal and tangent to an ellipse, major axis AB, minor axis CD.*

Draw the ellipse and mark a point *N* on the curve. As in Method 1, mark the position of the foci F^1 and F^2 on the major axis by using a radius of half the major axis from either *C* or *D*. Join F^1 and F^2 to point *N* and bisect the angle F^2NF^1. The bisector is the normal and, as in the circle, is perpendicular to its tangent. Draw the tangent perpendicular to the normal and to pass through *N*.

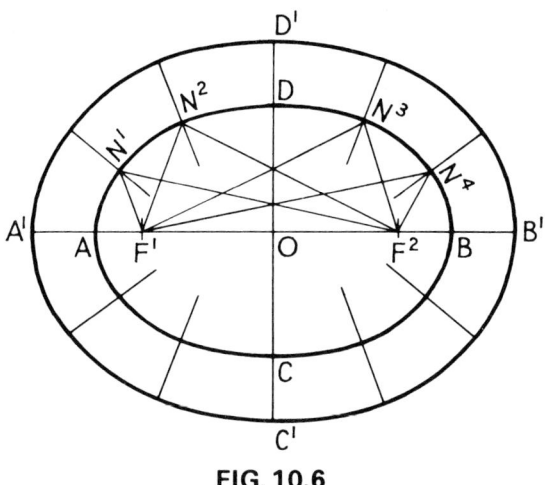

FIG. 10.6

FIG. 10.6 *To construct a curve, equidistant (distance AA') on the normals from an ellipse of major axis AB and minor axis CD.*

The required curve is $A'B'C'D'$ and could be either the outer curve of the border of an elliptical finger-plate or the outer curve of a path bordering an elliptical pond. Draw the ellipse $ABCD$ and produce the axes making DD', BB', and CC' each equal to AA'. As in Fig. 10.5, construct the foci F^1 and F^2, mark the points N^1, N^2, N^3, N^4, etc., on the curve and join each to F^1 and F^2. Bisect the angles at N^1, N^2, N^3, N^4 to construct the normals and produce these outside the ellipse. On each normal step off the distance AA' from the ellipse $ABCD$, and draw a smooth curve through these points.

EXERCISES

1. Draw ellipses by Methods 1, 2, and 3.
(i) Major axis 90mm, minor axis 60mm.
(ii) Major axis 98mm, minor axis 52mm.
   2. Draw ellipses by the trammel method.

(i) Major axis 75mm, minor axis 38mm.
(ii) Major axis 105mm, minor axis 45mm.
   3. Draw an ellipse, major axis 82mm, minor axis 60mm. (i) Show the foci, two normals and two tangents. (ii) Draw a curve about the ellipse at a distance of 12mm, measured along the normals.

CHAPTER 11

CURVES OF THE LOCUS OF A POINT

An introduction to the locus of a moving point has already been made in drawing an ellipse, but there are other much simpler loci such as that of a point moving directly from a fixed point *A* to a fixed point *B*; this is traced (on paper) by a pencil point and forms a straight line. Loci can be straight lines, curves of a definite type (e.g. circle, ellipse, cycloid), or other curves, sometimes more difficult, each one dependent upon the given conditions under which a point moves; yet all loci have their application in Machine Drawing, particularly in finding the amount of space required for the moving parts of a machine.

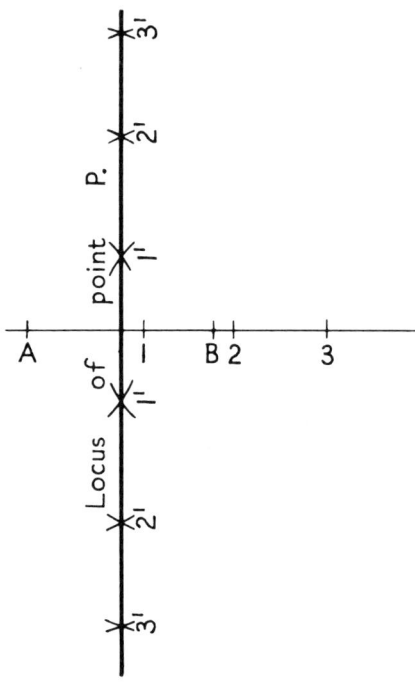

FIG. 11.1

FIG. 11.1 *To trace the locus of a point P moving at equal distances from and between two fixed points A and B.*

Draw a straight line and on it mark the points *A* and *B*. Mark point 1 on *AB* and at distance greater than $\frac{1}{2}AB$ from *A*. From point 1 on *AB* produced, step off two spaces of any convenient length, to give points 2 and 3. Since the conditions are that the distance from *P* to *A* must be equal to *P* to *B*, and that *P* must be between *A* and *B*, with centre *A* and radius *A*1 describe two arcs above and below *AB*. With centre *B* and the same radius, describe two arcs to cut the first two arcs at points 1′. By construction these points are equidistant from *A* and *B*. With *A* as centre and radius *A*2, describe two arcs above and below *AB* and, from *B* with the same radius, another two arcs to give the points 2′. Repeat from *A* and *B* with the radius *A*3 to give the points 3′. Draw a line through the six points 3′, 2′, 1′, 1′, 2′, and 3′, which is a straight line and the locus of *P*. This could be the locus of a centre point on the head of a piston moving inside a cylinder, the bottom of the cylinder walls being *A* and *B*.

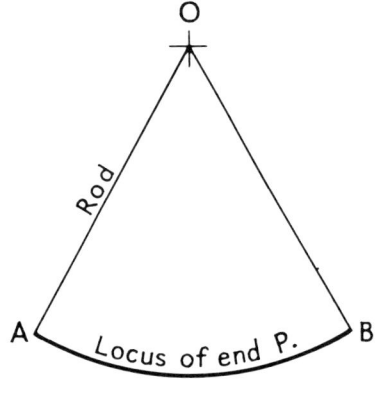

FIG. 11.2

FIG. 11.2 *To trace the locus of the end of a pendulum rod OP which swings about a fixed point O and between two fixed points A and B.*

Draw the points *A* and *B* and the rod *OP* in the position *OA*. Since the length of the rod is constant, the end *P* will trace an arc of radius *OP* and length *AB*. With *O* as centre and radius *OP*, draw the arc *AB* which is the locus of *P* and part of a circle.

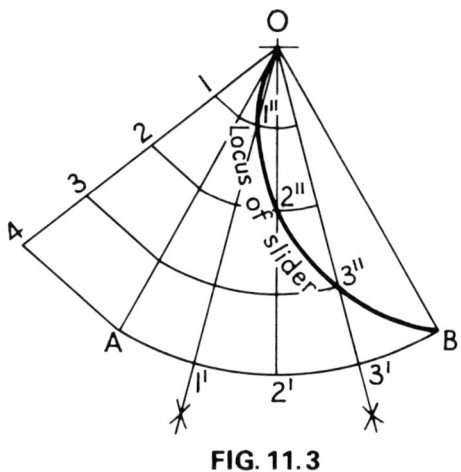

FIG. 11.3

FIG. 11.3 *To trace the locus of a point P on a slider which moves at a uniform rate from O to A, along the pendulum OA, during one complete swing from A to B.*

As in Fig. 11.2, trace the locus of the end of the pendulum *OA* to produce the arc *AB*. By means of a scale (as in Chapter 4) divide *OA* into four equal parts. Bisect the angle *AOB* and then the angles 2′*OA* and 2′*OB* and draw the radials 1′, 2′, and 3′. Since the slider moves down the pendulum at a uniform rate, when the pendulum has swung from *A* to 1′ (quarter of the arc) the slider will have descended from *O* to 1 (quarter of the length of of the pendulum), therefore with *O* as centre and radius *O*1 describe an arc to cut *O*1′ at 1″. Repeat with radii *O*2 and *O*3 to give the points 2″ and 3″ respectively. Draw a smooth curve, the locus of *P* , from *O* to *B* through 1″, 2″, and 3″.

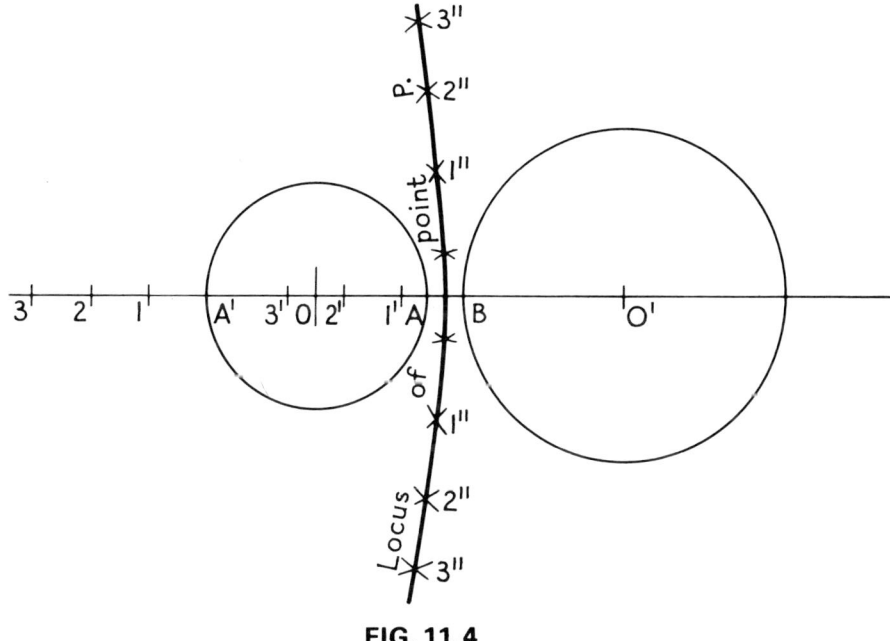

FIG. 11.4

FIG. 11.4 *To trace the locus of a point P which moves between and at an equal distance from the nearest point on the circumference of each of two circles at centres O and O′. The radii of the circles are OA and O′B, where O′B is greater than OA.*

Mark the centres *O* and *O′* and draw a straight line to pass through them. With centre *O* and radius *OA*, describe the smaller circle, and with centre *O′* and radius *O′B*, describe the larger circle. Bisect *AB* to give one position of *P*. On *OO′* and from *B*, step off three equal distances to give the points 1′, 2′, and 3′. On *OO′* produced and from *A′* step off a further three equal distances (i.e. *A′*1 equal to *B*1′) to give the points 1, 2, and 3. Since the distances of the points 1, 2, and 3 from the circumference of *OA* are equal to the distances of the points 1′, 2′, and 3′ from the circumference of *O′B*, then by using *O* and *O′* as centres and the distances from *O* to the points 1, 2, and 3 and *O′* to the points 1′, 2′, and 3′ as radii, the arcs described will intersect at points on the required locus. With *O* as centre and radii *O*1, *O*2, and *O*3 describe arcs below and above *OO′*. With *O′* as centre and radii *O*1′, *O*2′, and *O*3′ describe arcs, below and above *OO′*, to cut the first arcs at the points 1″, 2″, and 3″. Through 3″, 2″, 1″, mid-point of *AB*, 1″, 2″, and 3″ draw a smooth curve the locus of *P*.

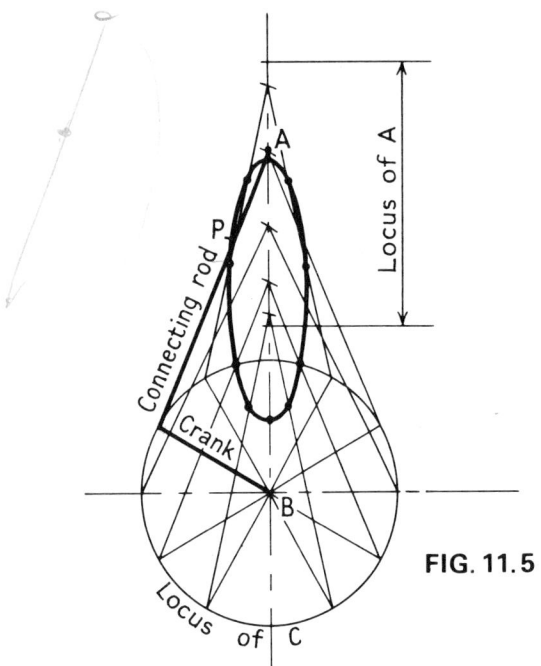

FIG. 11.5 *To trace the locus of a point P on a connecting rod AC, A moving along the straight line AB, and C hinged to a crank BC, moving about a fixed point B, e.g. to determine crank movement within the cylinder of an internal combustion engine.*

Draw *AB*, the centre line and line along which the point *A* moves. Draw *AC*, the connecting rod, and *BC*, the crank. With *B* as centre and radius *BC*, describe a circle and divide it into a number of equal parts to give the different positions of the crank. With length *AC*, step off from the divisions on the circumference on to line *AB* and join these points giving several different positions of the connecting rod. With length *CP* step off the different positions of the point *P* on the connecting rod. Draw a smooth curve through the points to give the required locus.

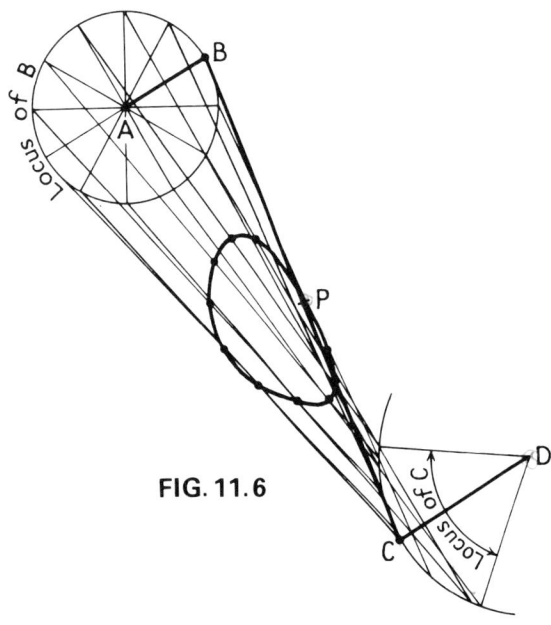

FIG. 11.6 *To trace the locus of a point P on BC, part of a system of three rods, AB, BC, and CD. A and D are fixed points of rotation, and rod BC is hinged at B and C.*

The linkwork could be part of a machine in which the amount of clearance, to allow free movement at point *P*, has to be found. Draw the rods *AB*, *BC*, and *CD*. With centres *A* and *D*, radii *AB* and *CD* respectively, describe two circles. Divide circle *AB* into a number of equal parts, to give the different positions of rod *AB* during its rotation, and from its circumference step off the length of rod *BC* on to the circumference of *CD*. Join these points to give the other positions of the rods and then from the circumference of *AB* step off the distance *BP* to mark the positions of *P*. Through the different points of *P* draw a smooth curve. The loci of the ends of all three rods are found and also the fact that *CD* cannot turn a complete circle.

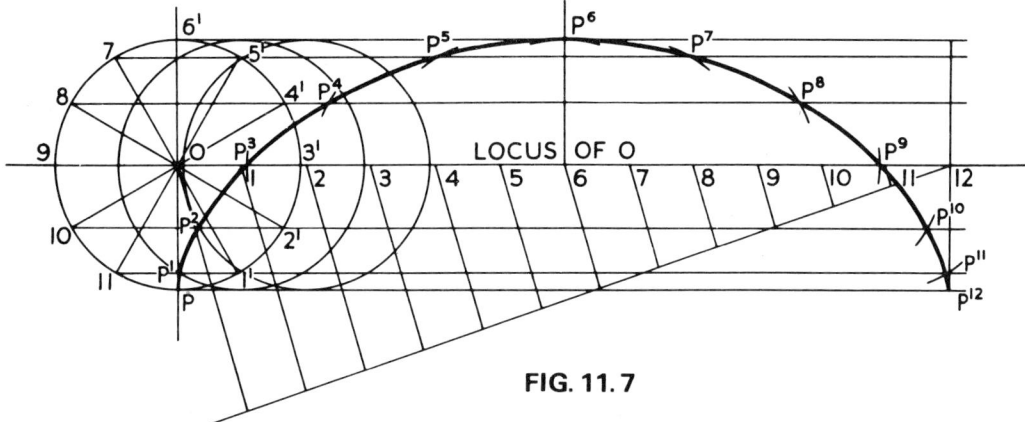

FIG. 11.7

Common Curves of the Locus of a Point

1. *Cycloid*

FIG. 11.7 *The locus of a point P on the circumference of a circle, radius OP, rolling along a straight line from P to P¹².*

To trace the locus. Draw the circle, radius *OP* and produce its horizontal centre line. Draw the straight line, PP^{12}, equal to the circumference of the circle and, from *O*, make the horizontal centre line the same length. Divide the centre line, *O* 12, and the circumference into 12 equal parts. From the divisions on the circumference of the circle draw parallels to PP^{12}. With *OP* as radius and centres 1, 2, 3, 4, 5, 7, 8, 9, 10, and 11, describe arcs to cut the parallel lines at $P^1, P^2, P^3, P^4, P^5, P^7, P^8, P^9, P^{10}$, and P^{11} (P^6 is vertically above 6). Draw a smooth curve through the points.

The figure shows the locus of the centre of the circle for one complete revolution and two other circles to show how *P* rises and moves forward as the circle rolls on PP^{12}. This construction is the base upon which epi-cycloids, hypo-cycloids, and trochoids can be drawn.

Cycloids have a practical use in constructing the shape of gear teeth, e.g. rack and pinion.

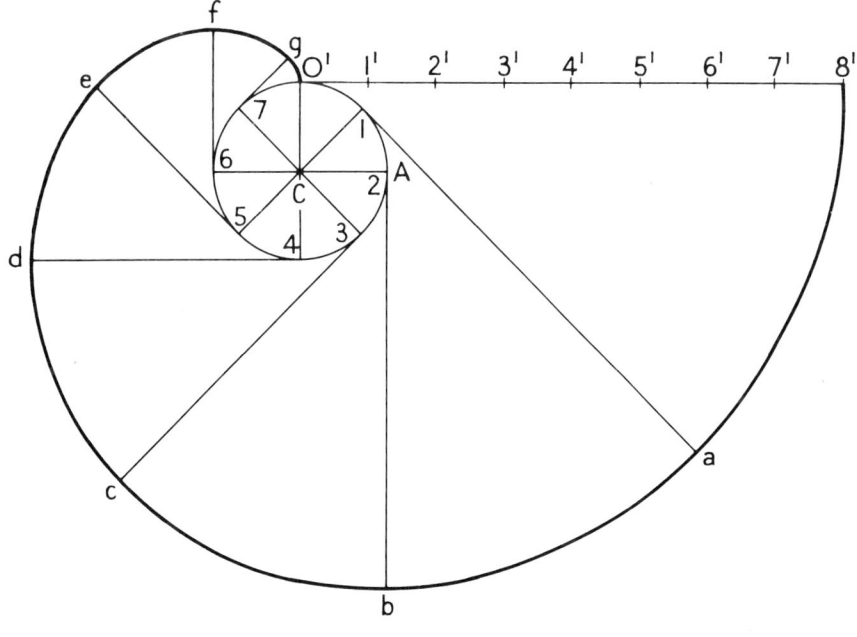

FIG. 11.8

2. *Involute of a Circle*

FIG. 11.8 *If one end of a flexible wire is fastened to a point on the circumference of a wheel, then the path traced by the free end of the wire, as it is wound round the wheel, is an involute of a circle.*

To trace the locus. Draw the circle centre C to represent the wheel and the tangent $0'8'$—equal in length to the circumference of the circle—to represent the wire fastened at point $0'$. Divide the circle into a number of equal parts and the tangent $0'8'$ into the same number, in this example eight. From each point of division on the circumference of the circle draw a tangent. Make the length of each tangent equal to the wire still unwound, i.e. at 1 to $1'8'$, at 2 to $2'8'$, at 3 to $3'8'$, at 4 to $4'8'$, at 5 to $5'8'$, at 6 to $6'8'$ and at 7 to $7'8'$, giving points a, b, c, d, e, f, g. From $8'$ and through $a, b, c, d,,e, f, g$ and 0, draw a smooth curve.

The involute curves are used in finding the profile of teeth of involute gears.

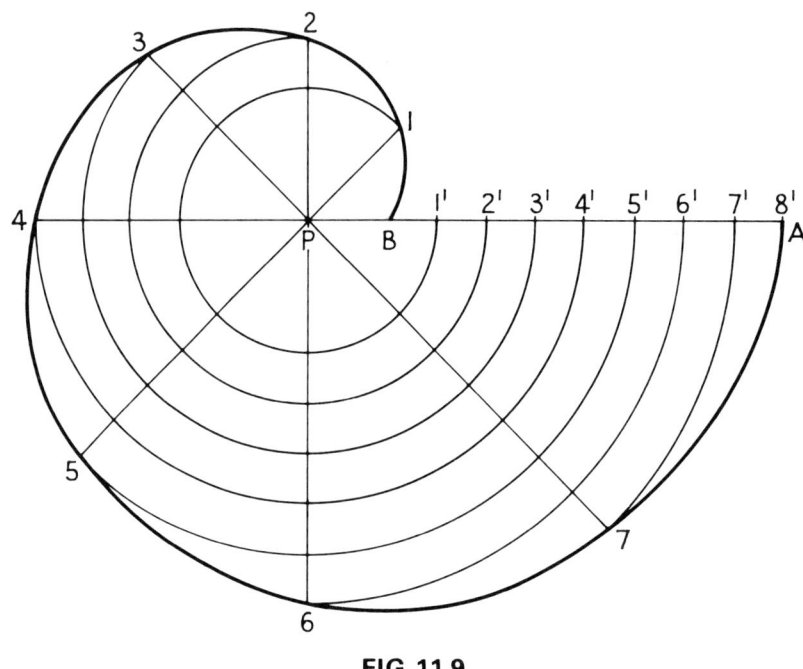

FIG. 11.9

3. *Archimedean Spiral*

FIG. 11.9 *This is the path of a point moving around and approaching a fixed point by equal amounts.*

To trace the locus. Mark the position of P, the fixed point or pole. From P draw the straight line PBA, where PA equals the farthest point position of the point from P, and PB the nearest position of the point to P. Divide AB into a number of equal parts and draw the same number of equally spaced radials from P, in this example eight. Since the radials are equally spaced the distance along each consecutive radial to the curve will be reduced in equal amounts; therefore, with P as centre and radii $P7'$ $P6'$, $P5'$, $P4'$, $P3'$, $P2'$, and $P1'$, describe arcs to cut the radials at 7, 6, 5, 4, 3, 2, and 1 respectively. Draw a smooth curve from 8 through these points to B.

In tracing a locus methods vary, and therefore to decide the nature and shape of a curve a simple system of card strips, joined by drawing pins, can be used generally indicating the method of making the trace.

4. *Conics*

The sectioning of a cone is described in Chapter 18, in which the parabola, hyperbola, and ellipse are true shapes projected from the cut or sectional surfaces of a cone. It is possible, but beyond the purpose of this book, to project from the cone the directrix and focus or foci of each curve, and it is on these that a general method for the construction of a conic, or conic section in plane geometry is derived.

The DIRECTRIX, DD, is a straight line from which perpendicular distances are measured to points on a conic curve and from which the axis AA (or centre line) of the conic is drawn. The FOCUS, F is a point on the axis from which distances are measured to points on the curve, such that for any point P there is the following ratio:

$$\frac{\text{distance from the focus to } P}{\text{perpendicular distance from the directrix to } P}$$

This ratio is known as the ECCENTRICITY, e.

For a parabola the eccentricity $=1$.

For a hyperbola the eccentricity is greater than 1.

For an ellipse the eccentricity is less than 1.

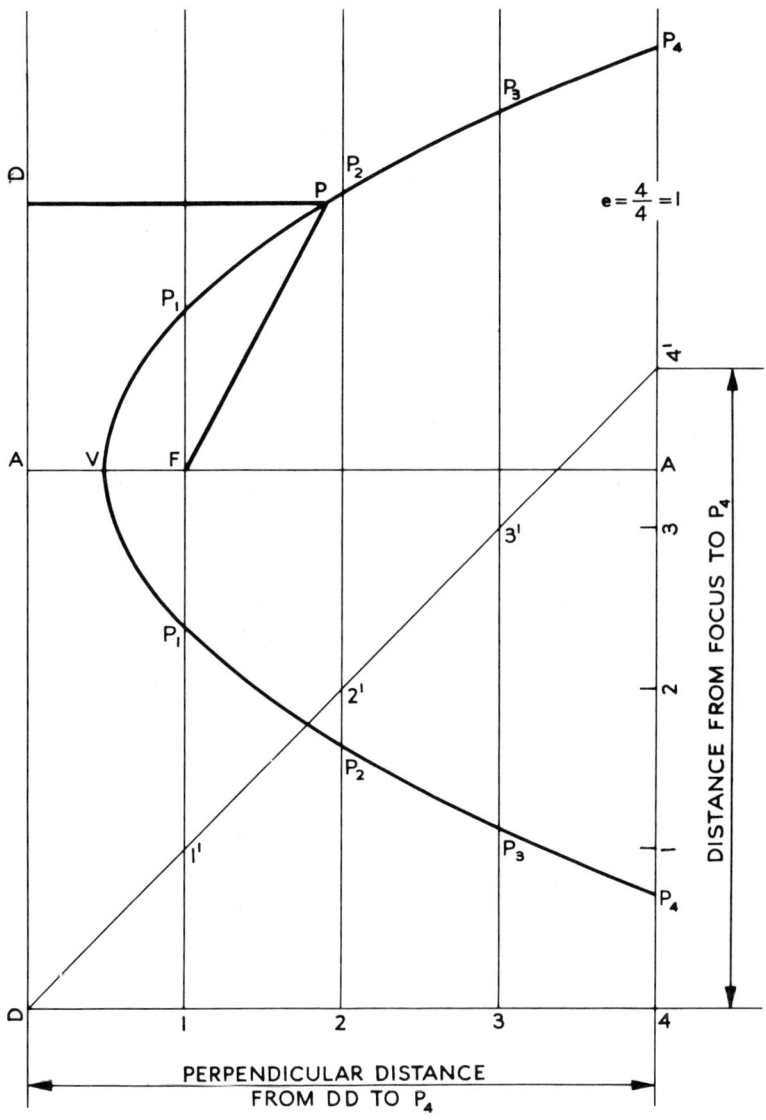

$$e = \frac{4}{4} = 1$$

FIG. 11.10

FIG. 11.10 *To construct a parabola by the general method, given the distance of the focus from the directrix.*

Draw the directrix, *DD*, in any convenient position and the axis, *AA*, perpendicularly to it. Mark the position of *F*. Since the eccentricity, *e*, of a parabola is 1, where the curve cuts the axis (known as the vertex, *V*), will be the midpoint between *DD* and *F*, such that *AV = VF*. Bisect *AF* to find *V*. Construct a right-angled triangle with one side perpendicular from *DD*, the other parallel to *DD*, and make the ratio of the lengths of the sides equal to the eccentricity, *e*, for a parabola = 1 and in this example 4 units each side. Draw parallels to *DD* through points 1, 2, and 3. With *F* as centre, length 11′, describe two arcs to cut 11′ produced below and above the axis to give points P_1. With *F* as centre and lengths 22′, 33′, and 44′ describe arcs on 22′, 33′, and 44′ produced respectively to give the remaining points on the curve, namely P_2, P_3, and P_4. Draw a smooth curve to pass through these points and *V*. $P_4 V P_4$ is the locus of point *P* and, under the given conditions, is a parabola.

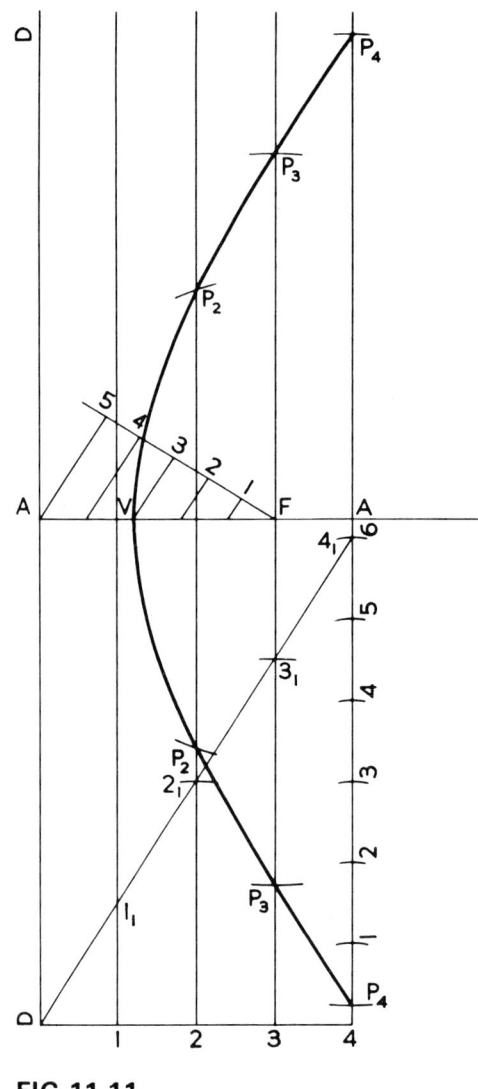

FIG. 11.11

FIG. 11.11 *To construct a hyperbola given the distance of F from DD and the eccentricity equal to $\frac{3}{2}$.*

Draw the directrix *DD*, the axis *AA* and the position of the focus, *F*. By means of a scale, divide *AF* in the ratio of 3 parts from *F* and 2 parts from $A(e = \frac{3}{2})$. As in Fig. 11.10, construct a right-angled triangle, but with the sides 2 and 3 units long, ratio equal to *e*. Divide the perpendicular from *DD* into 4 equal parts and the parallel to *DD* into 6 equal parts. Complete the right-angled triangle, produce parallels to *DD* through the points 1, 2, and 3 and with *F* as centre and lengths 11′, 22′, 33′, and 44′ describe the points on the curve. Draw a smooth curve through $P_4 V P_4$ to form the hyperbola.

The ellipse can be drawn by the same method, but this is unnecessary at this stage since the methods described in Chapter 10 are more easily understood.

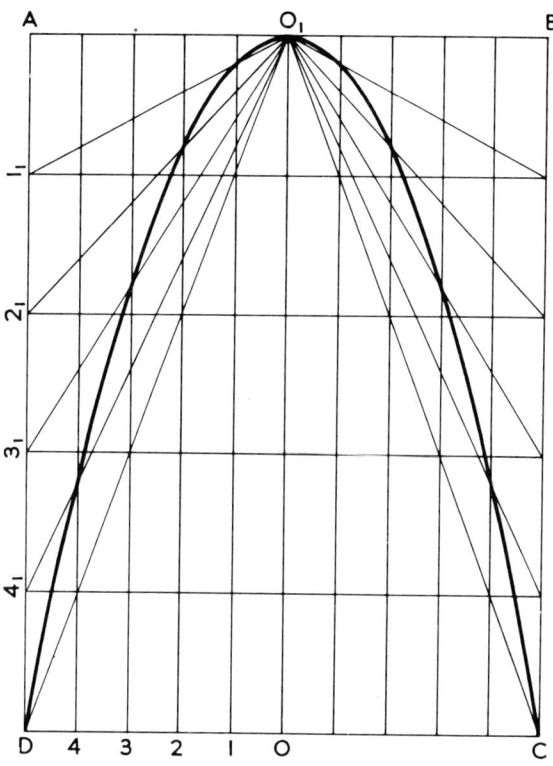

FIG. 11.12

FIG. 11.12 *To construct a parabola within a given rectangle, ABCD.*

Draw the given rectangle *ABCD*. Bisect *AB* and draw the centre line OO_1. Divide *AD* and *DO* into the same number of equal parts, in this example 5. Draw parallels to *AD* through 1, 2, 3, and 4 and join 1_1, 2_1, 3_1, and 4_1 to O_1. Draw a smooth curve through the points of intersection and repeat the construction in the half O_1BCO to complete the parabola. This method is a simple construction of the path of a projectile, e.g. a cricket ball, when it is thrown into the air, neglecting the factors which normally affect its course.

Conic curves, whether constructed by projection from cones or as the loci of points, are of more than mathematical value as they form the shapes of many common objects, e.g. reflectors, and rarer objects, in particular the large aerials used to transmit and receive signals from outer space.

EXERCISES

1. *A* and *B* are two fixed points 45mm apart. Trace the locus of a point *P*, which is equidistant from both *A* and *B*, and moves between them.

2. The length of a pendulum is 75mm and in one complete swing it moves through an angle of 60°. A slider moves from the top of the pendulum to the bottom at a uniform rate during one complete swing. Represent the slider by a point *P* and trace its locus for one complete swing of the pendulum.

3. Two circles are of 30mm and 45mm radii and their centres are 105mm apart. Trace the locus of a point *P* which moves between and at an equal distance from the nearest point on the circumference of each circle.

4. A crank *AB* is 45mm long, moves about a fixed point at *A*, and is hinged at *B* to a rod *BC*, 75mm long. End *C* moves along a straight line. Draw the locus of a point on *BC*, 37mm from *B*.

5. In a similar system to Exercise 4, *AB* is 37mm *BC* is 90mm, and the point is 52mm from *B*. Trace the locus of the point.

6. Three rods *AB*, *BC*, and *CD* are hinged at *B* and *C*. *A* and *D* are fixed points of rotation 120mm apart. *AB* is 60mm, *BC* is 112mm, and *CD* is 45mm. A point *P* on *BC* is 37mm from *C*. Draw the locus of *P* when the system moves about *A* and *D*.

7. The circle of a cycloid curve is (i) 60mm diameter, (ii) 52mm diameter. Trace the cycloid in each case.

8. A piece of flexible wire is equal in length to the circumference of a wheel to which one end is fixed. Trace the involute when the wheel is (i) 45mm diameter, (ii) 38mm diameter.

9. A point moves round and approaches a fixed point *P* by equal amounts. If the distances of the point from *P* are: (i) farthest 120mm, nearest 22mm, (ii) farthest 150mm, nearest 37mm, trace the spiral for each.

10. Construct a parabola when the focus is: (a) 45mm from the directrix, and (b) is 60mm from the directrix.

11. Construct a hyperbola (a) when the eccentricity is $\frac{3}{2}$ and the focus 75mm from the directrix, and (b) when the eccentricity is $\frac{5}{3}$ and the focus 60mm from the directrix.

12. Construct a parabola within a rectangle (a) 150mm by 100mm, and (b) 180mm by 135mm.

A. PROBLEMS INVOLVING PRACTICAL APPLICATIONS

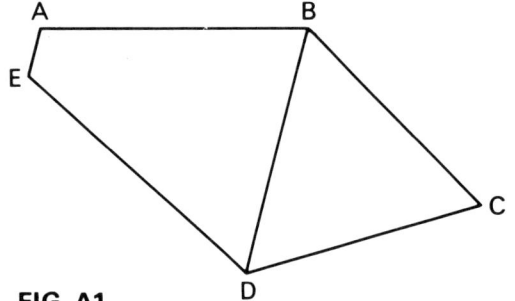

FIG. A1

Set 1.

1. Fig. A1 represents the frame of a bicycle. Construct the shape without using set squares or a protractor to the following dimensions:
$AB = 146$mm; $BD = 135$mm; $BC = 127$mm; $CD = 135$mm; $AE = 26$mm; angle $ABD = 75°$ and angle $BAE = 105°$.

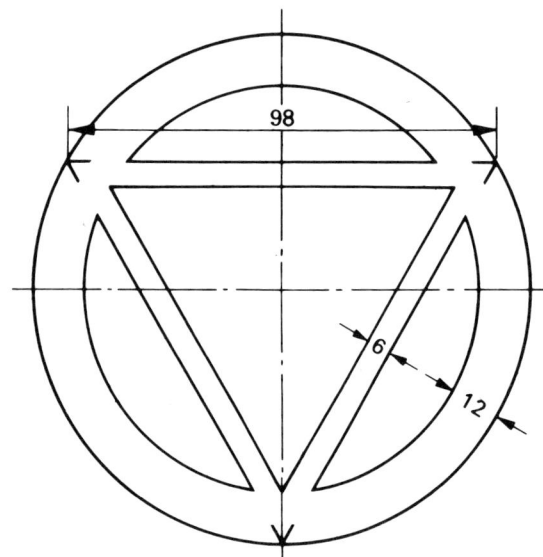

FIG. A2

2. Fig. A2 represents the outline of a road traffic Stop Sign of which the triangle is equilateral. Construct the view.

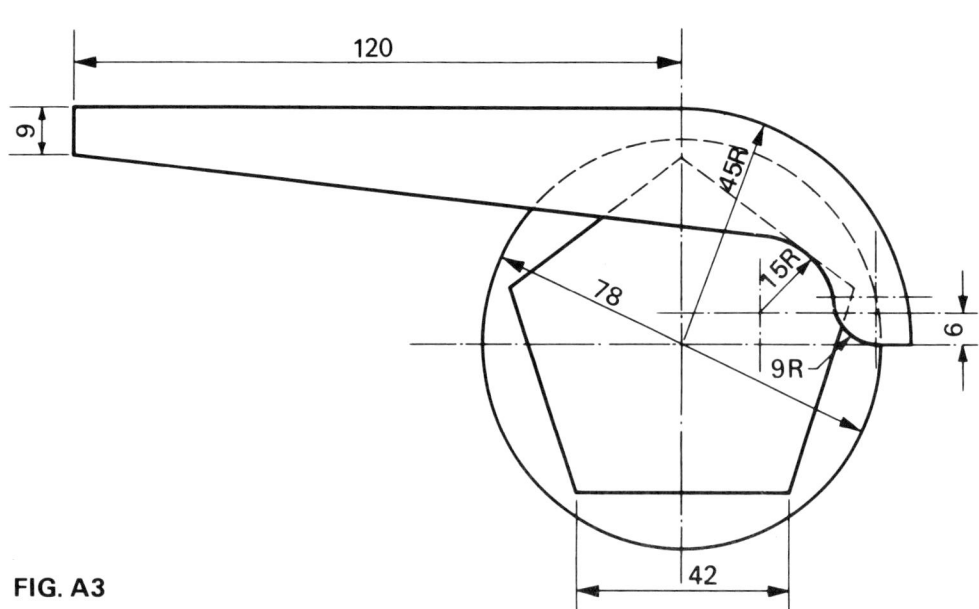

FIG. A3

3. Fig. A3 represents the shape of a bicycle chain guard and part of the crank wheel. Construct the view and clearly show the construction of the tangent to the 15mm radius arc.

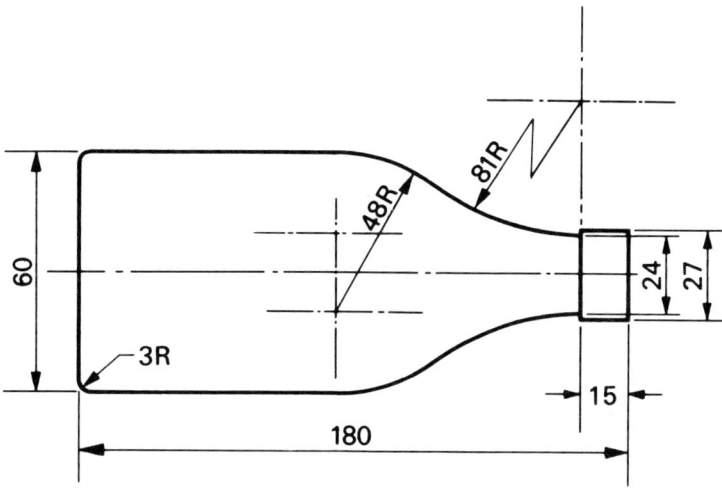

FIG. A4

4. Fig. A4. is the profile of a die used for casting mineral water bottles. Construct the profile and clearly indicate all normals and centres.

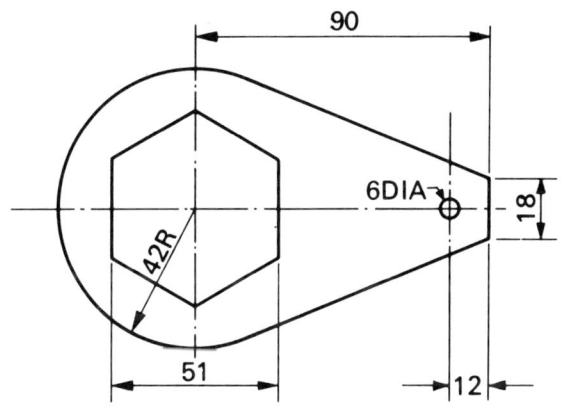

FIG. A5

Set 2.

5. Fig. A5 represents the shape of a locking plate. Construct the shape without the use of set squares or a protractor and draw the normals to the two tangents.

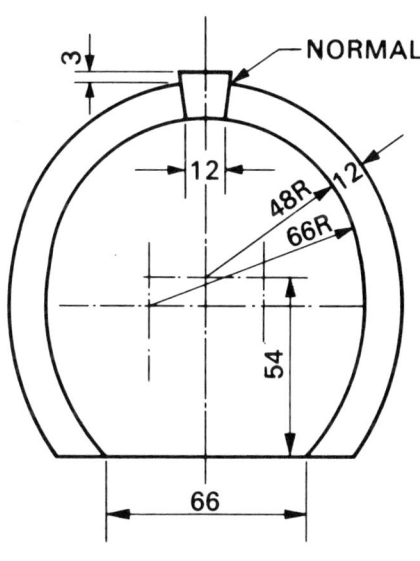

FIG. A6

6. Fig. A6 represents the section of a tunnel. Construct the shape and show all normals.

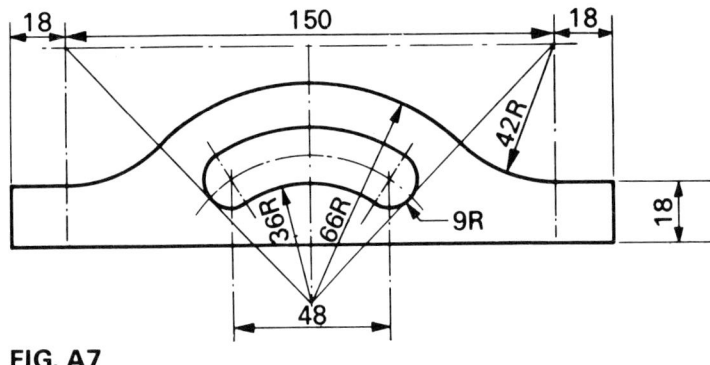

FIG. A7

7. Fig. A7 represents the end of a tray. Construct the view and clearly indicate the common normals.

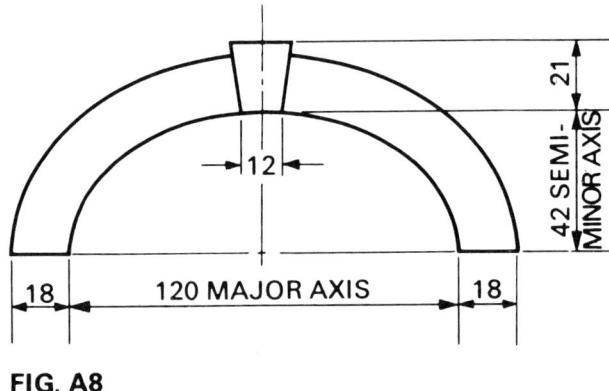

FIG. A8

8. Fig. A8 represents a semi-elliptical arch. Construct the view and show all the construction of the outer equidistant curve.

SOLID GEOMETRICAL DRAWING

Solid Geometrical Drawing consists of drawing views of the different faces of an object by the projection of points and lines on to planes of projection. It is a means of showing the exact size, and shape, in three dimensions and is therefore the basis of all Machine and Structural Drawings used in industry. The most common projection used is the Orthographic (or Orthogonal) projection, in which lines are projected from the object on to two or more planes, the projectors to any one plane are parallel, and all are perpendicular to the plane.

CHAPTER 12

PROJECTION OF POINTS AND STRAIGHT LINES

Figure 12.1 shows four boards hinged together and opened to form four quadrants each having one Vertical Plane and one Horizontal Plane (V.P. and H.P.). Only one quadrant is used for any one drawing and this is usually the first, but the third quadrant can be used. (First and Third Angle are British Standard Methods). Suspended in the first quadrant is a rectangular block from which projectors have been drawn on to the V.P. and H.P. (known as the Principal or Co-ordinate planes of projection). In Fig. 12.2 the H.P. has been folded back, along the line *XY*, to make one plane surface with the V.P., and the views shown, together with any additional ones, form an orthographic projection of the block in First Angle Projection. The line *XY* is used to indicate the fold between the principal planes, and, in first angle, above the *XY* line is the V.P. and below it the H.P.

Since the principles of orthographic projection depend on the projection of points and lines to make a view, then the obvious approach is to learn how to project points and lines. The solid can be represented by a large pin, the head of pencil point diameter and the stem a straight line of pencil point cross-section.

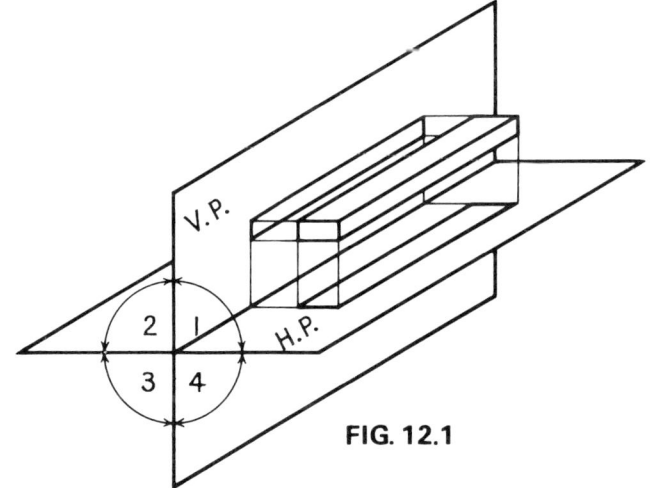

FIG. 12.1

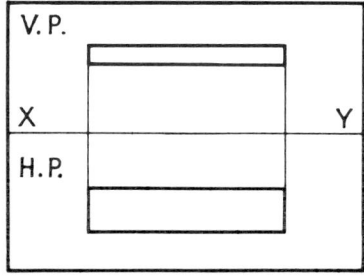

FIG. 12.2

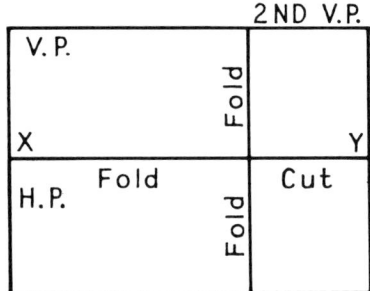

FIG. 12.3

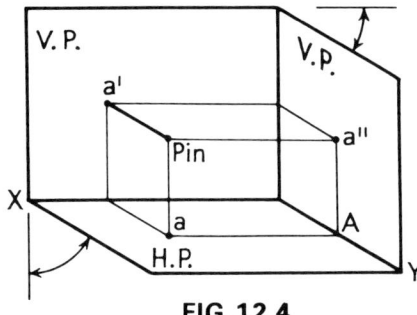

FIG. 12.4

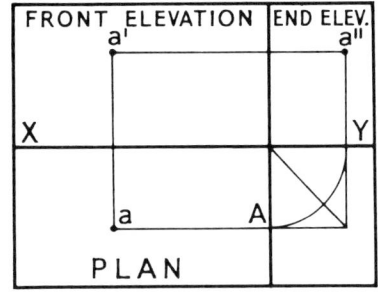

FIG. 12.5

FIG. 12.6

FIGS. 12.3, 12.4, and 12.5 *To make an ortho-graphic projection of a point.*

Fold and cut a small sheet of drawing paper (as shown in Figs. 12.3 and 12.4) to make the principal planes and a second V.P. Using these paper planes, stick a pin perpendicularly into the V.P., then the stem of the pin is the projector of the head to the V.P., point a'. If a projector from the head perpendicular to the H.P. could be drawn, then this would enter the H.P. at point a and if a projector from the head perpendicular to the second V.P. could be drawn, then this would enter the second at a''. Mark these three positions of the pin-head on the planes and join the points with projectors. Hold the pencil point at A, the foot of the vertical projector on the second V.P., and turn the plane back moving the pencil with it and so drawing a projector. Fold the H.P. flat to form a plane surface with the two V.P.s.

On the flat sheet of paper, Fig. 12.5, there are three views of the pinhead, one in each V.P., namely the Front Elevation and End Elevation, and one on the H.P., the Plan. The Front Elevation and Plan are on the principal planes and are essential views, but the End Elevation is on an auxiliary plane and is not always necessary. Notice that: (i) the projector from the H.P. to the second V.P. is a quadrant, but can be replaced with a much neater method which consists of drawing a square and its diagonal, (ii) the views seen from the front of the pin are placed behind it. This is very important when drawing an end elevation.

FIG. 12.6 *To make an orthographic projection of a line AB which has one end on the V.P., is perpendicular to the V.P., is parallel to and above the H.P.*

Use the paper planes and a pin to determine the position of the line. The point of the pin, end A of the line, will be stuck in the V.P. and end B will be the end seen and shown as the Front Elevation, b'. The stem of the pin will be perpendicular to the V.P. and so projected on to the H.P. to give the Plan, ab. The second V.P. can be placed in any convenient position on the Principal Planes, and the stem of the pin, seen from the left, will be projected on to this plane as a horizontal line to give the End Elevation $a''b''$. Having considered the positions of the three views, mark the position of point b' in the V.P. and draw projectors from b' to the H.P. and the second V.P. Mark the points ab on the H.P. equal to the length of the line, and draw a projector from b to the second V.P. The actual views are now 'lined in' (darkened) and the projectors left to show the Geometrical Construction.

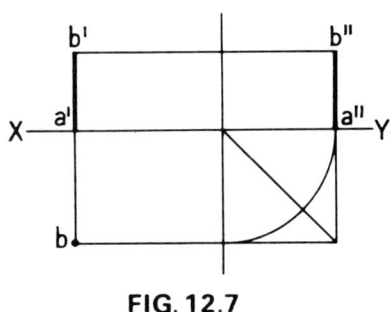

FIG. 12.7

FIG. 12.7 *To make an orthographic projection of a line AB which is in front of the V.P., is perpendicular to and on the H.P.*

Use the paper planes and pin to determine the position of the line and fix the position of the second V.P. The point *A* will be in the H.P. and end *B* will be seen; therefore mark the position of *b* on the H.P. and draw projectors from *b* to the V.P. and the second V.P. Mark the points $a'b'$ on the V.P. equal to the length of the line, and draw a projector from b' to the second V.P. Line in the views b, $a'b'$, and $a''b''$.

Using the pin and planes, draw the following projections:

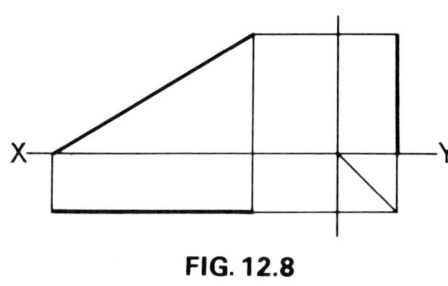

FIG. 12.8

FIG. 12.8 *The end of a 90mm straight line is on the H.P., inclined at an angle of 30° to the H.P., is parallel to and 22mm in front of the V.P.*

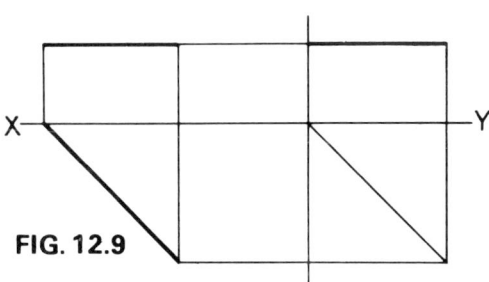

FIG. 12.9

FIG. 12.9 *The end of a 75mm straight line is on the V.P., inclined at an angle of 45° to the V.P., is parallel to and 30mm above the H.P.*

True Length, Angle of Inclination, Vertical and Horizontal Traces

It is often necessary to find the true length of a line, its angles of inclination to the planes and the points, where, if it were produced, it would meet the V.P. and the H.P. These points are termed the Vertical and Horizontal Traces. In the previous exercises the true length is given in the elevation or plan, but when a line is inclined to both the V.P. and the H.P., then it is necessary to introduce an auxiliary plane to find the true length, inclination, and traces. The true or actual lengths must be found in the construction of metal or wooden-framed roofs, particularly the hips which are inclined to the V.P. and the H.P.

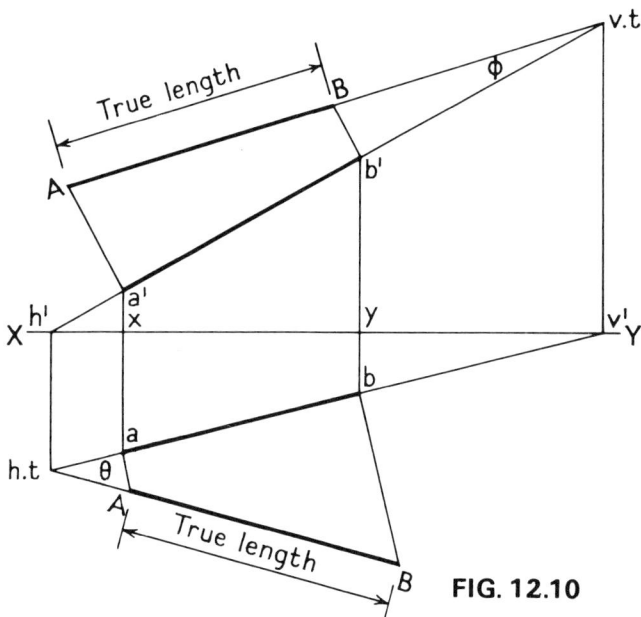

FIG. 12.10

Fig. 12.10 *To find the true length, angles of inclination to both planes, and the vertical and horizontal traces of a straight line AB. The elevation $a'b'$ is 105mm long, a' is 15mm above and b' is 67mm above the H.P. In the plan ab, the point b is 22mm in front of the V.P. and the point a is 45mm in front of the V.P.*

Draw the elevation $a'b'$ and from it project the plan ab. Using ab as the position of an auxiliary vertical plane on the H.P., draw projectors from ab respectively equal to the projectors $a'x$ and $b'y$ in the V.P., giving points A and B. Join AB and produce to meet ab produced. Because ab is the base of an auxiliary vertical plane, AB produced is on this plane. Aa is equal to $a'x$ and Bb is equal to $b'y$. Then AB is the true length of the line, θ is its angle of inclination to the H.P., and the point h.t. (where the line enters the H.P.) is its horizontal trace.

If a similar method is used, making the elevation $a'b'$ the position of an auxiliary perpendicular plane on the V.P., then by making Aa' equal to ax and Bb' equal to by, AB again is the true length of the line, ϕ is its angle of inclination to the V.P., and v.t. is its vertical trace. In all cases the projectors are perpendiculars to the planes.

The traces can be found by producing the elevation and plan to the line of interesection of the principal planes and then drawing projectors from these points. Produce ab to v', and draw a projector from v' to $a'b'$ produced to give the point v.t. Produce $a'b'$ to h', and draw a projector from h' to ab produced to give the point h.t.

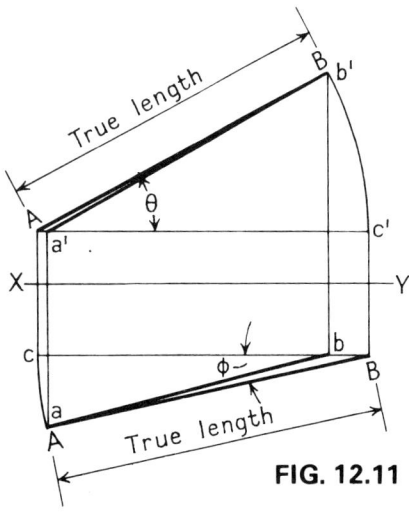

FIG. 12.11

Fig. 12.11 *A second method to find the true length and angles of inclination of a straight line.*

This method is that of rabatment (lines were rabatted in Figs. 7.11 and 7.12) of the plan and elevation of the line, parallel to the *XY* line, and then projecting the new points. From a' and b, draw lines parallel to the *XY*. With a' as centre and radius $a'b'$, describe an arc to give point c', making $a'c'$ equal to $a'b'$. Project c' on to cb produced to give point B, then because $a'c'$ is the true length of the elevation, AB must be the true length of the line and ϕ its inclination to the V.P. With b as centre and radius ab, describe an arc to give point c and project to the V.P. to give point A. By the same reasoning, AB is the true length of the line and θ its angle of inclination to the H.P.

Compare the results of both methods.

EXERCISES

1. Draw the front and end elevations and plan of a point, (i) 45mm above the H.P., 30mm in front of the V.P.; (ii) 67mm above the H.P., 15mm in front of the V.P.

2. Draw the front and end elevations, plan, and traces of a straight line: (i) 105mm long, inclined at 60° to the V.P., one end on the V.P., 15mm above and parallel to the H.P. (ii) 82mm long, inclined at 40° to the H.P., one end 7mm above the H.P., 22mm in front of and parallel to the V.P.

3. The elevation of a line is $a'b'$ and is 82mm long, a' is 15mm and b' is 75mm above the H.P. The plan of the line is ab, b is 15mm in front of the V.P. and a is 45mm in front of the V.P. Find:
(i) The true length.
(ii) The angles of inclination.
(iii) The vertical and horizontal traces.
(iv) Repeat (i) and (ii) by the first method. Cut on AB, aA, bB, $a'A$, and $b'B$. Fold at right angles on ab, XY, and $a'b'$.

4. A line 112mm long is inclined at 30° to the H.P. and 50° to the V.P. Draw its projections.

5. The plan of a line AB is 98mm long and inclined at 45° to the V.P. The true angle of inclination to the H.P. is 30°. Draw the plan and elevation of the line and show its horizontal trace.

6. The plan of a line ab is 75mm, is inclined to the V.P. at 45°, end a is 15mm in front of the V.P. and b the farther end. The h.t. of the line is on the XY and end B of the line is 90mm above the H.P. Draw its plan, elevation, and find its angle of inclination to the H.P. and its true length.

CHAPTER 13

TRACES OF PLANES

In addition to finding the true length of a line, as in Chapter 12, auxiliary planes are also used to determine the true shape, and, in sections, to expose hidden detail of machine parts. These planes have to be inserted in any part of an object which is being drawn. Projectors have to be drawn on to them and their positions known on the principal planes. The positions are called the Horizontal and Vertical Traces and it is on to these traces that auxiliary views are projected.

In Chapter 12 the vertical and horizontal traces of a line were shown as points in the V.P. and H.P.

respectively; but a plane, which has a greater width than a line, will make a straight cut in one or both of the principal planes; therefore a Vertical Trace of a Plane (v.t.) is *the line on which the auxiliary plane cuts the V.P.,* and a Horizontal Trace of a Plane (h.t.) is *the line on which the auxiliary plane cuts the H.P.* To understand fully the meaning of traces, a flat card or piece of tinplate should be inserted into cuts in the principal planes: these can be of card or drawing paper.

Types of Planes

Each figure shows a sketch of the auxiliary plane in the principal planes and the trace or traces.

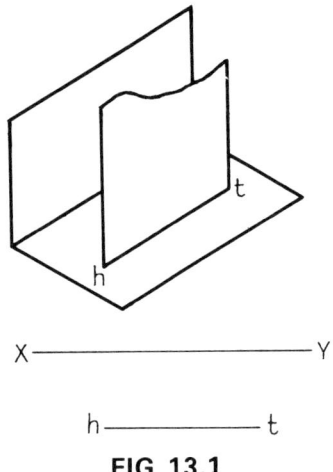

FIG. 13.1 *Vertical and parallel to the V.P. The horizontal trace is parallel to the V.P. No vertical trace.*

FIG. 13.1

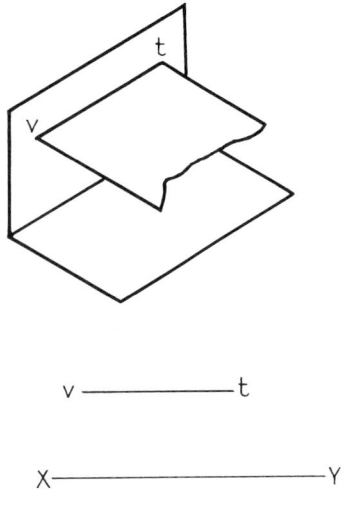

FIG. 13.2 *Perpendicular to the V.P. and parallel to the H.P. The vertical trace is parallel to the H.P. No horizontal trace.*

FIG. 13.2

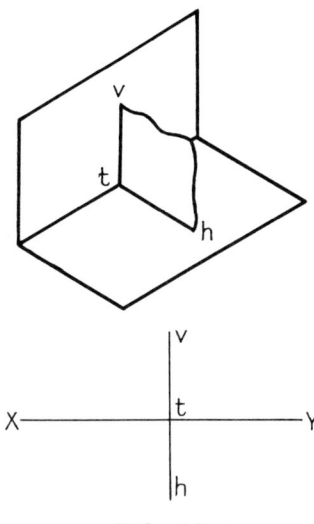

FIG. 13.3

FIG. 13.3 *Vertical and perpendicular to the V.P. The vertical trace is perpendicular to the H.P. and the horizontal trace perpendicular to the V.P. This plane is the second V.P. already introduced in Chapter* 12.

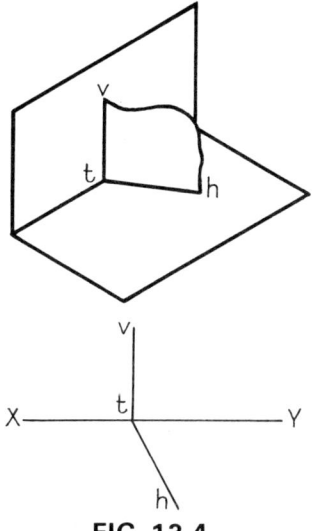

FIG. 13.4

FIG. 13.4 *Vertical and inclined to the V.P. The vertical trace is perpendicular to the H.P. and the horizontal trace inclined to the V.P.*

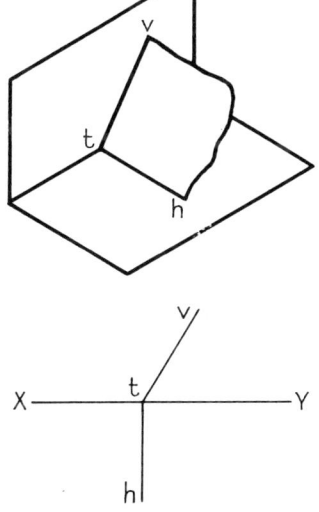

FIG. 13.5

FIG. 13.5 *Perpendicular to the V.P. and inclined to the H.P. The vertical trace is inclined to the H.P. and the horizontal trace perpendicular to the V.P.*

The planes in Figs. 13.1 to 13.5 are all Perpendicular Planes.

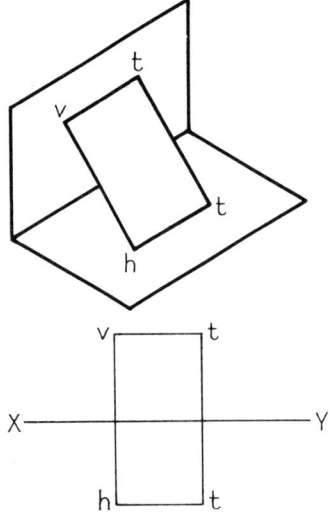

FIG. 13.6

FIG. 13.6 *Oblique. Inclined to both the H.P. and V.P. The vertical and horizontal traces are parallel to the H.P. and V.P. respectively.*

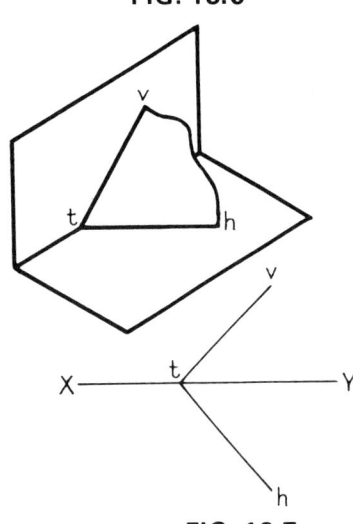

FIG. 13.7

FIG. 13.7 *Oblique. Inclined to both the H.P. and V.P. The vertical trace is inclined to the H.P. and the horizontal trace inclined to the V.P.*

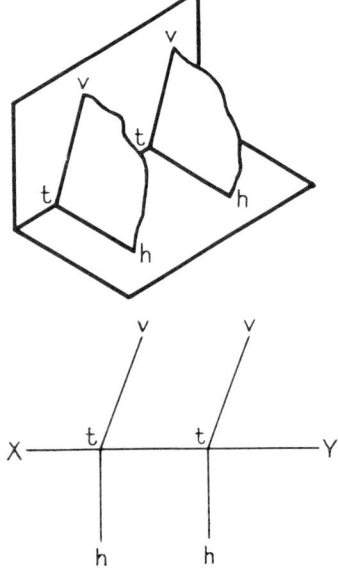

FIG. 13.8

FIG. 13.8 *Parallel perpendicular planes inclined to the H.P. The traces are parallel and the same as those in Fig. 13.5.*

Because a point is a position on a solid object, then it must be known how to draw projectors from a point (in space) to a plane and hence find its perpendicular distance from the plane.

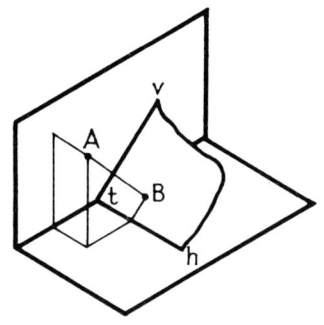

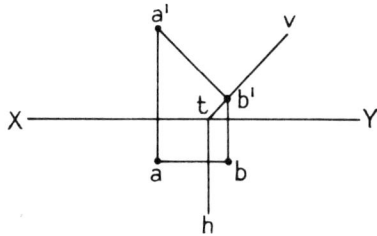

FIG. 13.9

FIG. 13.9 *To find the perpendicular distance of a point A from a perpendicular plane given the traces of the plane and the position of point A.*

The sketch shows point *A* and its perpendicular *AB* to the plane. Draw the traces of the plane, v.t. and h.t. As in the previous chapter, draw and project the elevation and plan of point *A*, giving a' and a. Draw a perpendicular from the v.t. to the elevation a' and from the h.t. to the plan a. The perpendicular on the V.P., $a'b'$, is the distance of point *A* from the perpendicular plane (true length of *AB*). Project *b*, on the H.P., from b' and the perpendicular from a, then ab is the plan of *AB*.

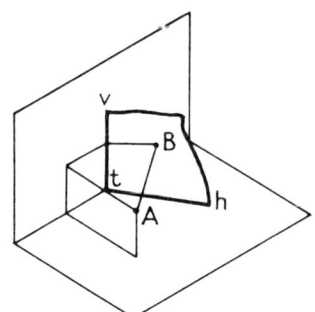

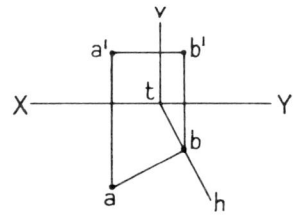

FIG. 13.10

FIG. 13.10 *To find the perpendicular distance of a point A from a perpendicular plane given the traces of the plane and the position of point A.*

The sketch shows point *A* and its perpendicular *AB* to the plane. In Fig. 13.9 the auxiliary plane is inclined to the H.P., but in this problem it is inclined to the V.P. and it will be necessary to draw the plan of *A* and *AB* and then project to the V.P. Draw and project the plan and elevation of the point *A* giving a and a'. Draw perpendiculars from the h.t. and the v.t. to a and a' respectively. The perpendicular ab is the distance of point *A* from the plane, and by projecting point *b* to the V.P., $a'b'$ is the front elevation of *AB*.

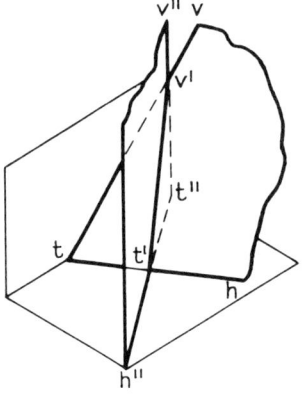

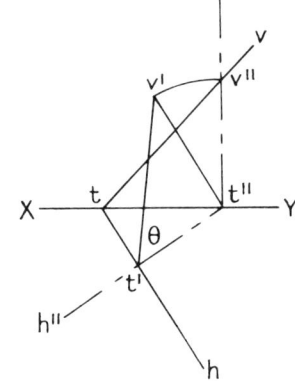

FIG. 13.11

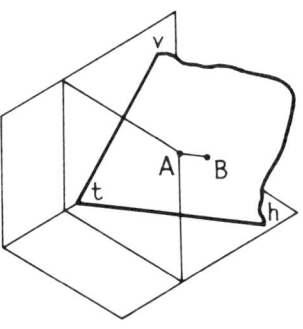

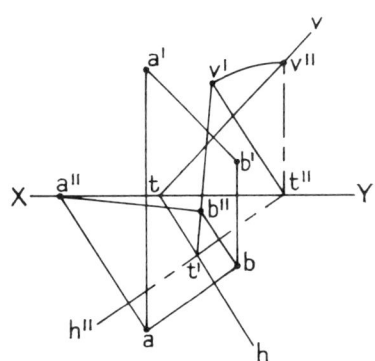

FIG. 13.12

FIG. 13.11 *To transform an oblique plane vth into a perpendicular plane v′t′h.*

This example is one in which a second plane cuts an auxiliary plane and is useful when true angles have to be found of odd-shaped constructions, such as metal hoppers and roofs. By projecting the traces of a vertical (perpendicular to the H.P.) plane $v''t''h''$ on to the traces of the oblique plane vth, the oblique plane is transformed into a perpendicular plane which gives the true angle of inclination to the H.P. Draw the traces of the oblique plane, vt and ht, and, at any convenient point on ht, draw the horizontal trace of the vertical plane, $h''t''$, perpendicular to ht. Draw the vertical trace, $v''t''$, and rabat the trace to $v't''$ perpendicular to $h''t''$. Join v' to the point where $h''t''$ cuts the h.t. of the oblique plane, t', then $v't'$ and ht' are the traces of the transformed plane. $v't'$ is the true length on the auxiliary vertical plane to the V.P. and θ the true angle of inclination to the H.P. The trace, ht', of the transformed plane is on the trace, ht, of the oblique plane.

FIG. 13.12 *To draw the perpendiculars and find the perpendicular distance of a point A from an oblique plane vth, given the traces of the plane and the position of the point A.*

The oblique plane can be transformed into a perpendicular plane by the method of Fig. 13.11; then the problem is that of projecting a point on to a perpendicular plane, as in Fig. 13.9, and from this to the oblique plane. Draw the traces of the oblique plane, vt and ht, and the elevation, a', and plan, a, of the point A. Transform the oblique plane into a perpendicular plane, as in Fig. 13.11, its traces being $v't'$ and ht'. Draw the elevation of point A to the perpendicular plane by projecting from a, perpendicular to $t''h''$, and making the distance of a'' (the elevation) above $t''h''$ equal to a' above tt''. Since a'' and a are the projections of A to the transformed plane, then following the method of Fig. 13.9, draw a perpendicular from a'' to $v't'$ to give point b'', and project from b'', on to a perpendicular from a to ht', to give point b. $a''b''$ is the perpendicular distance of A from the plane. Because the horizontal traces of the oblique and transformed planes are the same, then ab is the plan of the perpendicular to both planes, therefore a projector from b will be b' on the oblique plane. Draw a projector from b, and make the distance of b, above tt'' equal to that of b'' above $h''t''$. Join $a'b'$ to complete the elevation of the perpendicular to the oblique plane.

In transforming an oblique plane the auxiliary plane replaces the V.P.

EXERCISES

1. Draw the traces of the following auxiliary planes:

(i) A vertical plane, parallel to and 45mm in front of the V.P.

(ii) A horizontal plane 60mm above the H.P.

(iii) A plane perpendicular to the H.P. and V.P.

(iv) A plane inclined at an angle of 30° to the H.P. and perpendicular to the V.P.

(v) A plane inclined at an angle of 45° to the V.P. and perpendicular to the H.P.

(vi) Two parallel planes, 82mm apart, each inclined 60° to the H.P. and perpendicular to the V.P.

(vii) A plane obliquely inclined at 30° to the H.P. and 45° to the V.P.

(viii) A plane obliquely inclined at 40° to the H.P. and 60° to the V.P.

2. Draw the perpendicular projectors from a point to a plane:

(i) A perpendicular plane inclined at 40° to the H.P., the point 60mm above the H.P. and 45mm in front of the V.P. and its projector 15mm in front of the horizontal trace.

(ii) A vertical plane inclined at 50° to the V.P., the point 67mm above the H.P. and 60mm in front of the V.P. and its projector 22mm in front of the vertical trace.

3. Repeat Exercise 2 (i) and (ii) for a point 52mm above the H.P., 75mm in front of the V.P., and its projector 38mm in front of the point of intersection of the traces on XY.

4. An oblique plane is inclined at 30° to the H.P. and 45° to the V.P. Draw its traces, transform it into a perpendicular plane, and indicate its true inclination to the H.P.

5. Repeat Exercise 4 for an oblique plane inclined at 40° to the V.P. and 30° to the H.P.

6. A point A is 60mm above the H.P. and 60mm in front of the V.P. Its projector is 15mm behind the point of intersection of the traces (i.e. it passes through them) of an oblique plane:

(i) Using the oblique plane of Exercise 4, draw its projections and find its perpendicular distance from the plane.

(ii) Draw its projections and find its perpendicular distance from an oblique plane inclined at 20° to the V.P. and 50° to the H.P.

CHAPTER 14

PROJECTION OF SIMPLE PLANE FIGURES (LAMINAE) AND AUXILIARY PROJECTIONS

The plane figures (or laminae) used here introduce the measuring of two dimensions with only the third dimension of pencil-point thickness. These figures can be real solids when they are cut out of thin sheet-metal or card, as in a Machine Drawing in which a thin washer, or one lamination of the core of an electric motor, is drawn.

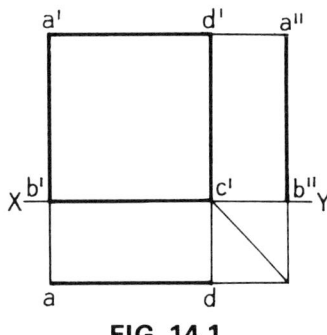

FIG. 14.1

FIG. 14.1 *To draw the plan and front and end elevations of a square lamina, ABCD, with edge BC on the H.P., parallel to and in front of the V.P., and the surface ABCD vertical.*

Using the principles of Chapter 12, the front elevation will be a square, the end elevation and plan both straight lines. Draw the square $a'b'c'd'$ on the V.P. (by the method given in Fig. 5.9), $b'c'$ on the XY line, and produce the sides as projectors to the H.P. and the second V.P. Draw the plan ad parallel to the XY line and between the projectors on the H.P. Draw a projector from ad to the second V.P., and on it, and between its other projectors, the end elevation $a''b''$. Line in the three views.

Figure 14.1 is the projection of a lamina in a common position, and so the views are true in length and shape, but in actual practice a surface to be projected may not always be parallel to one of the principal planes, and this causes a distortion in one or more of the views. The laminae in Figs. 14.2, 14.3, and 14.4 are projected in the same way as that in Fig. 14.1 but are inclined to one of the principal planes.

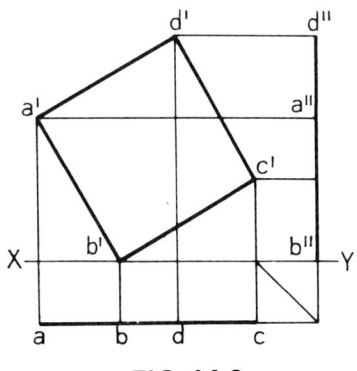

FIG. 14.2

FIG. 14.2 *To draw the plan and front and end elevations of a square lamina, ABCD, with corner B on the H.P., edge BC inclined at 30° to the H.P. parallel to and in front of the V.P., and its surface vertical.*

The front elevation is a square and both the plan and end elevation are straight lines. These only give the thickness of the lamina, and not the length of one side as in Fig. 14.1. Draw the square, $a'b'c'd'$, on the V.P., with b' on the XY and $b'c'$ inclined at 30° to the XY line. By the method used in previous examples, project the plan, adc, the end elevation, $b''a''d''$, and line in the views.

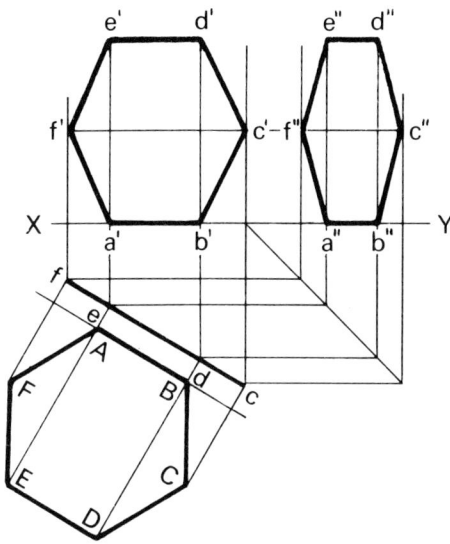

FIG. 14.3

FIG. 14.4

FIG. 14.3 *To draw the plan and front and end elevations of a regular hexagonal lamina, ABCDEF, with edge AB on the H.P. and inclined at 30° to the V.P., and its surface vertical. A is in front of and nearer to the V.P. than B.*

The front and end elevations are not regular hexagons, the plan is a line inclined to the V.P.; therefore a separate and additional view of a regular hexagon must first be drawn, and from it the length of the plan and vertical dimensions of the elevations are taken. Draw the regular hexagon, *ABCDEF*, in a convenient position (if possible so that dimensions can be projected from it). Produce side *AB* and, perpendicular to it, draw lines from the points *E, D, C,* and *F*. Since these lines can be produced to cut the plan and then projected from it on to the V.P. and the second V.P.—because they are perpendicular and therefore parallel—they are a means of transferring lengths from the true figure to other views. Lines of this kind, used for the transference of of lengths on a figure, are generally called *Ordinates.* Between the ordinates from *F* and *C*, draw the plan *fedc* on the H.P. and inclined at 30° to the *XY*. Since *f, e, d,* and *c* are the same distances apart as the ordinates through *FEDC*, draw projectors from these points to the V.P. and the second V.P. The projectors in the V.P.s are ordinates of the figures to be drawn. Therefore make *a'e'* equal to *AE* and *f'* above *a'b'* produced equal to *F* above *ED* or *AB* produced. Draw horizontal projectors from *e'* and *f'* which give the remaining points on the elevations. Join these points and line in the three views.

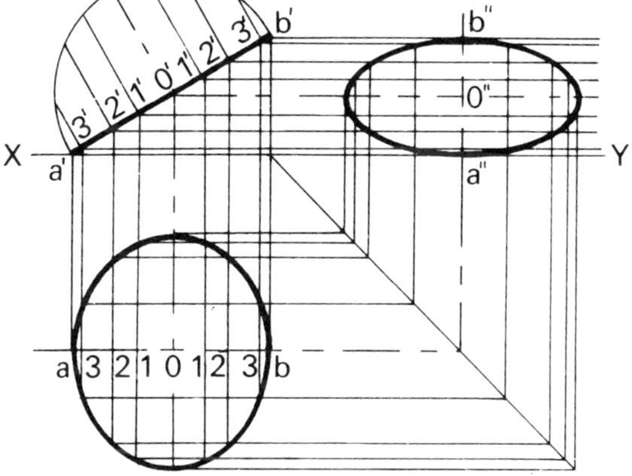

FIG. 14.4 *To draw the plan and front and end elevations of a circular lamina, OA, with one point on the H.P., its surface inclined at 30° to the H.P., the front elevation a line inclined at 30° to the H.P., and the lamina in front of the V.P.*

The front elevation is a straight line inclined at 30° to the H.P., but the end elevation and plan are ellipses and an additional view of a circle must be drawn. Draw the front elevation, *a'o'b', a'* on *XY*, equal to the diameter of the lamina and on it draw a semicircle to represent the additional view. On *a'o'b'*, and on each side of the centre *o'*, step off equal distances to mark the points *1', 2',* and *3'*, erect ordinates at each of these points and project them and the centre line to the H.P. and the second V.P. Since circles are drawn about centre lines, transfer the lengths of each ordinate on the semicircle to the corresponding ordinates, both sides of the centre line, on the H.P. Draw a smooth curve through the points to complete the plan. Draw projectors from these points to the second V.P. and, through the intersections of the projectors, draw the end elevation.

Projection of Auxiliary Elevations and Auxiliary Plans

Auxiliary elevations and plans are projected on to auxiliary planes. The main use of auxiliary elevations is that of showing the true shapes of objects which cannot be projected on to the principal planes.

Important points:

(i) An auxiliary elevation is projected from a plan and an auxiliary plan is projected from an elevation (on the principal planes, a plan is projected from one or both of the elevations, and elevations are pro-

jected from the plan).

(ii) A horizontal trace is required for an auxiliary elevation, a vertical trace for an auxiliary plan— Chapter 13 has dealt with traces, which are often termed auxiliary *XY* lines, but in this book are shown as either h.t. or v.t.

(iii) All projectors are perpendicular to the trace and parallel to the direction of a given arrow. An arrow shows the direction of viewing the object. Alternatively, the position of the auxiliary plane is given.

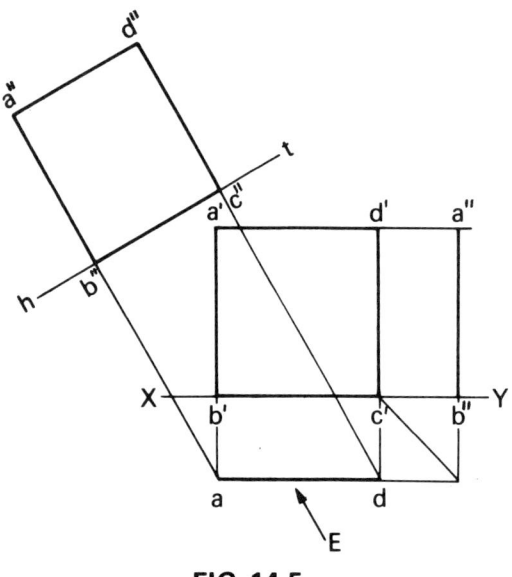

FIG. 14·5

FIG. 14.5 *To draw the plan, front and end elevations, and an auxiliary elevation, looking in the direction of the arrow E, of a square lamina ABCD, with edge BC on the H.P., in front of and parallel to the V.P., and its surface vertical.*

As in Fig. 14.1, draw the plan and front and end elevations. Since an auxiliary elevation is to be drawn, it will be on a vertical plane inclined to the H.P. and perpendicular to the arrow *E*. For convenience, draw the h.t. above the front elevation and projectors to it from *a* and *d* on the plan. The auxiliary elevation will be the same vertical height above its trace as that of the front or end elevations above *XY* (it is on a vertical plane), so transfer the distances $a'b'$ to $a''b''$ and $c'd'$ to $c''d''$. Complete the auxiliary elevation by joining $a''d''$.

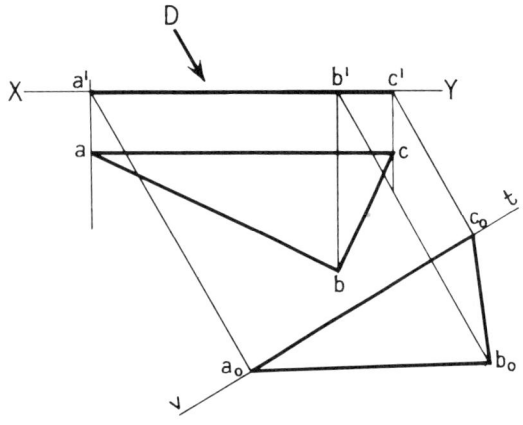

FIG. 14.6

FIG. 14.6 *To draw the plan, front elevation, and an auxiliary plan, looking in the direction of the arrow D, of a triangular lamina ABC lying on the H.P. Edge AC is parallel to and in front of the V.P.*

Draw the plan and from it project the front elevation. The auxiliary plane, for a plan, will be inclined to the H.P. and its v.t. perpendicular to the direction of the arrow. Draw the trace in any convenient position and projectors to it from a', b', and c' on the front elevation. Transfer the perpendicular distance of *b* above *ac* in plan to b_0 above a_0c_0 (above the v.t.). Join the points a_0, b_0 and c_0 to complete the auxiliary plan.

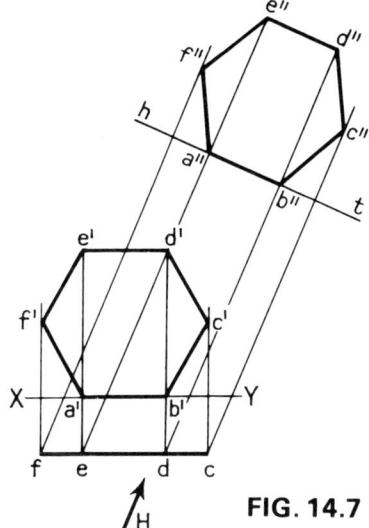

FIG. 14.7

FIG. 14.7 *To draw a plan, front elevation, and an auxiliary elevation, looking in the direction of arrow H, of a regular hexagonal lamina ABCDEF. Edge AB is on the H.P., parallel to and in front of the V.P., and its surface is vertical.*

Draw the front elevation and plan. From the plan, draw projectors parallel to arrow *H* (as in Fig. 14.5); draw the h.t. and transfer dimensions, above the *XY*, from the front elevation to the projectors, and above the h.t. Join *a″*, *b″*, *c″*, *d″*, *e″*, and *f″* to complete the auxiliary elevation.

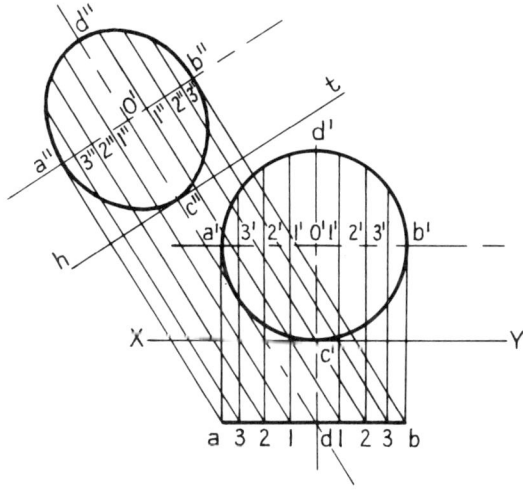

FIG. 14.8

FIG. 14.8 *To draw the plan, front elevation, and an auxiliary elevation, on a vertical plane inclined at 32° to the V.P., of a circular lamina OA. One point is on the H.P. and its surface is in front of and parallel to the V.P.*

Draw the front elevation and project the plan. Using a similar method to that of Fig. 14.4, step off equal distances each side of the centre line on the plan (points 1, 2, and 3) and draw projectors from these to the front elevation. Draw the h.t. inclined at an angle of 32° to *XY* and projectors from the plan perpendicular to the h.t. Since it is a circular lamina, draw a centre line distance *OA* above the h.t. and, as in Fig 14.4, transfer the distances, at points $1'$, $2'$, and $3'$, on the front elevation, from the centre line to the circumference, to each side of the centre line above the h.t., at points $1″$, $2″$, and $3″$. Draw a smooth curve to complete the auxiliary elevation.

FIG. 14.9 *To draw a plan, front elevation, and an auxiliary plan, on a perpendicular plane inclined at 45° to the H.P., of a circular lamina OA lying on the H.P. and in front of the V.P.*

The only difference in this method from that of Fig. 14.8 is that projectors are drawn from the front elevation and dimensions are transferred from the plan. Draw the plan and project the front elevation. Step off equal spaces each side of the centre line, points $1'$, $2'$, and $3'$, on the front elevation and draw projectors from these to the plan. Draw the v.t. at an angle of 45° to *XY*, draw projectors from the front elevation perpendicular to the v.t. and a centre line distance *OA* below the v.t. Transfer the lengths from the points 1, 2, and 3 on the plan to each side of the centre line below the v.t. and complete the auxiliary plan by drawing a smooth curve.

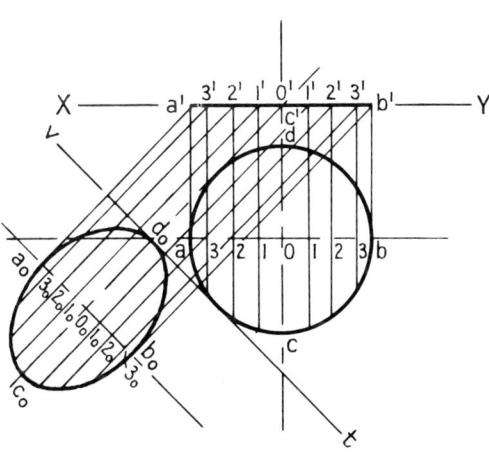

FIG. 14.9

EXERCISES

1. Draw a front elevation, end elevation and plan of each of the following:

(i) A square lamina of 37mm side, one corner on the H.P. and an edge at 30° to the H.P. Its surface is vertical, parallel to and 15mm in front of the V.P.

(ii) A rectangular lamina of 90mm and 45mm sides, one short edge on the H.P. and at right angles to the V.P. Its surface is inclined at 60° to the H.P. and the distance between the lamina and the V.P. is 15mm.

(iii) A triangular lamina, isosceles in shape, with 45mm base and an altitude of 67mm. The base of the lamina on the H.P., 15mm from and inclined at 45° to the V.P., and its surface vertical.

(iv) A regular hexagonal lamina of 30mm side, one edge on the H.P. and inclined at 30° to the V.P. Its surface is vertical and its centre 60mm in front of the V.P.

(v) A circular lamina of 38mm radius with its surface inclined at 30° to the H.P. One point of the circumference rests on the H.P., its centre is 60mm in front of the V.P., and the front elevation is a line inclined at 30° to the H.P.

2. Using the dimensions of the laminae given in Exercise 1, draw:

(i) A front elevation, plan, and an auxiliary elevation on a vertical plane, inclined at 45° to the V.P., of a square lamina with its surface vertical, parallel to and 15mm in front of the V.P. and one edge on the H.P.

(ii) A front elevation, plan and an auxiliary plan, looking in the direction of an arrow inclined at 30° to the H.P., of an isosceles triangular lamina. Its surface is horizontal, 22mm above the H.P. and one long edge is on the V.P.

(iii) A plan, front elevation, and an auxiliary elevation, looking in a direction of 40° to the V.P., of a regular hexagonal lamina with one edge on the H.P., 30mm in front of and parallel to the V.P. and its surface is vertical.

(iv) A front elevation, plan, and an auxiliary plan, on a perpendicular plane inclined at 50° to the H.P. of a circular lamina lying on the H.P. with its centre 60mm in front of the V.P.

CHAPTER 15

PROJECTION OF RECTILINEAL SOLIDS

The word solid, in this chapter, referes to objects having three dimensions all of which are greater than pencil-point size and so can be measured with a ruler. These are common solids, built up from common plane figures, and are the basis of the greater number of machine or building drawings, e.g. bolts, bearings, girders, concrete lintels, etc.

Cube, Right Prism, and Right Pyramid

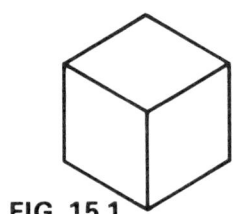

FIG. 15.1

FIG. 15.1 *A cube is a solid bounded by six equal squares.*

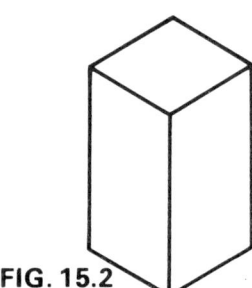

FIG. 15.2

FIG. 15.2 *A prism is a solid having its end faces equal, similar, and parallel and all its side faces parallelograms. Its axis is a line joining the centres of the two ends and the prism is right when this is a perpendicular to the ends. Prisms are named from the shape of their ends.*

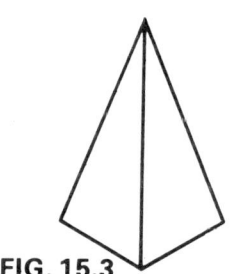

FIG. 15.3

FIG. 15.3 *A pyramid is a solid with triangular side faces, which meet at a point called the vertex, and a base, which is a plane figure. Pyramids are named from the shape of their bases. The axis is a line joining the vertex to the centre of the base. When a pyramid is right, the axis is perpendicular to the base.*

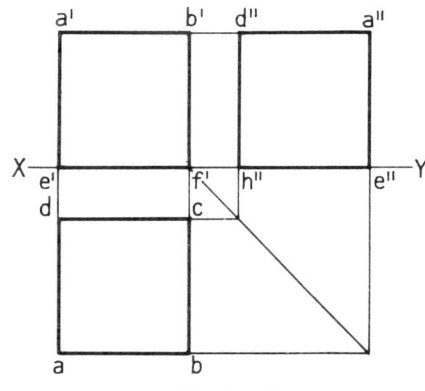

FIG. 15.4

FIG. 15.4 *To draw the plan and front and end elevations of a cube of side AB on the H.P., with surface CDHG in front of and parallel to the V.P.*

The three views will all be squares, and the method used is that of Chapters 12 and 14, of projecting views on to the principal and second vertical planes. Draw the front elevation, $a'b'f'e'$, on the V.P., $e'f'$ on XY, and projectors from it to the H.P. and the second V.P. Between the projectors on the H.P. and parallel to the XY draw the plan, *abcd*. Draw projectors from the plan to the second V.P. and through their inter-sections with projectors from the front elevation, draw the end elevation, $a''e''h''d''$. It is important to notice that in First Angle Projection, as stated in Chapter 12, the view seen is projected behind the object and this is indicated by the lettering.

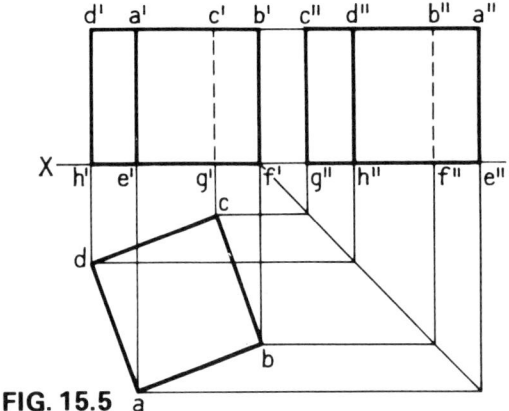

FIG. 15.5

FIG. 15.5 *To draw the plan and front and end elevations of a cube of side AB on the H.P., its surface CDHG in front of and inclined at* 20° *to the V.P.*

The difference between this projection and Fig. 15.4 is that already explained in Figs. 14.2, 14.3, and 14.4. In this example the plan is inclined to the V.P. but is a true square—the front and end elevations are only true in height and not in length. Draw the plan, *abcd*, and from it draw projectors

to the *XY* and the second V.P. Make the distance of *d'b'* above the H.P. equal to side *AB* and produce *d'b'* as a projector to the second V.P. Complete the views by lining in, and in particular indicating the edges not seen by *Hidden Detail Lines*—lines consisting of short dashes, *g'c'* and *f"b"*.

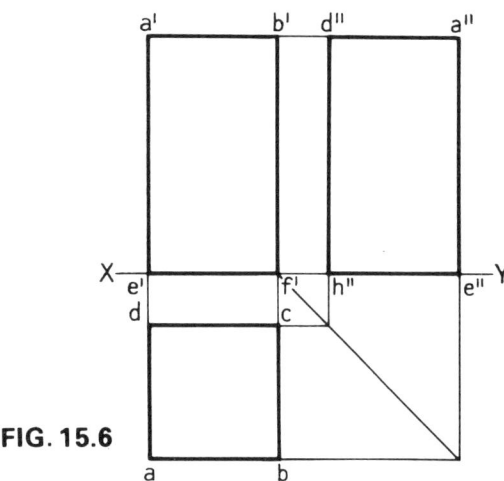

FIG. 15.6

FIG. 15.6 *To draw the plan and front and end elevations of a right square prism of end AB and length AE, with one end on the H.P. and side CDHG in front of and parallel to the V.P.*

The front and end elevations will both be rectangles and the plan a square. Draw the front elevation, *a'b'f'e'*, on the V.P., *e'f'* on *XY*, and from it project the plan and end elevation, *abcd* and *a"e"h"d"* respectively, by the method used in Figs. 15.4 and 15.5.

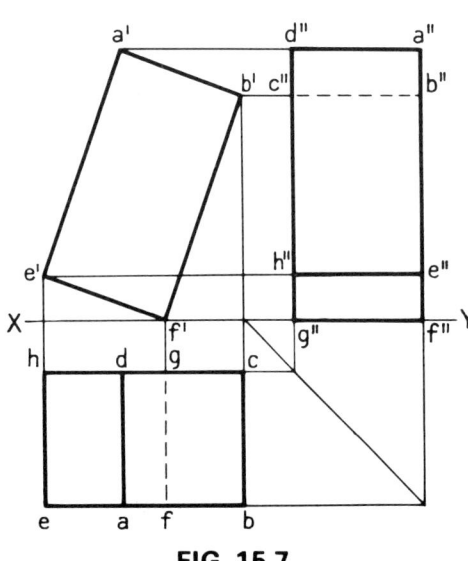

FIG. 15.7

FIG. 15.7 *To draw a plan and front and end elevations of a right square prism of end AB and length AE, with edge FG on the H.P., in front of and perpendicular to the V.P., and end EFGH inclined at* 20° *to the H.P.*

Only the front elevation will be a true side (a rectangle of sides *AE* and *AB*). The plan and end elevation will be rectangles, each consisting of two surfaces of the prism, similar to the front and end elevations of Fig. 15.5. Draw the front elevation, *a'e'f'b'*, on the V.P. with *f'* on the *XY* and *e'f'* at 20° to the *XY*. Draw projectors from the front elevation to the H.P. and the second V.P.; draw the plan between the projectors on the H.P. and projectors from it to the second V.P. Lining in will complete the three views. Hidden detail lines are again drawn, in the lining in, to indicate the hidden edges.

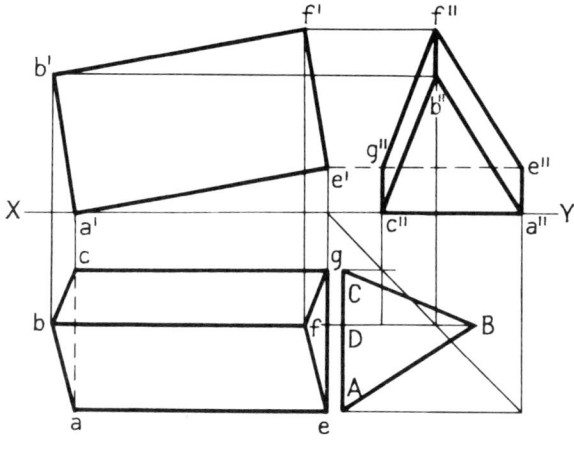

FIG. 15.8

FIG. 15.8 *To draw the plan and front and end elevations of a right triangular prism of end ABC and length AE. Edge AC is on the H.P. and perpendicular to the V.P. Side AEGC is inclined at 10° to the H.P. and the prism is in front of the V.P.*

This is a similar problem to that of the hexagonal lamina in Fig. 14.3. A separate and additional view of the triangle must be drawn before any one of the three views can be projected. Draw the additional view, *ABC*, of the triangle on the H.P. and from it projectors parallel to the V.P. *BD* is an ordinate and the perpendicular height of an end of the prism, therefore on the V.P. draw the front elevation making $a'e'$ equal to the length of the prism and inclined at 10° to the *XY*, $a'b'$ and $e'f'$ equal to *BD*. Draw projectors from the front elevation to the H.P. and the second V.P. and complete the plan, *abcefg*. From the plan draw projectors to the second V.P., and through the intersections of these with the projectors from the front elevation, draw the end elevation, $a''b''c''e''f''g''$. Line in the hidden detail and full lines.

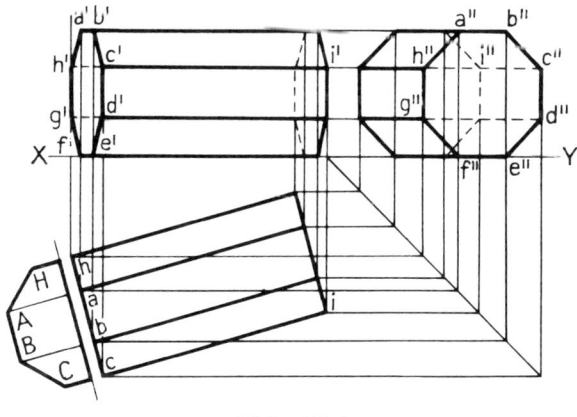

FIG. 15.9

FIG. 15.9 *To draw the plan and front and end elevations of a right octagonal prism of end ABCDEFGH and lenght AI. One face is on the H.P. in front of the V.P. and its long edges inclined at 15° to the V.P.*

This is another projection which requires an additional view, but in this example only half the octagon, *ABCH*, has been drawn. Draw the additional view on the H.P., inclined at 15° to the *XY* and in position such that the plan can be projected from it. Draw ordinates from *A* and *B* and produce these, and the sides at *C* and *H*, as projectors. Draw the plan, *abchi*, between these projectors and from it draw projectors to the V.P. and the second V.P. Draw the front elevation, $a'b'c'd'e'f'g'h'i'$, on the V.P. and from it project the end elevation, $a''b''c''d''e''f''g''h''i''$, as in the preceding examples, by joining the points of intersection of the projectors. Line in the views.

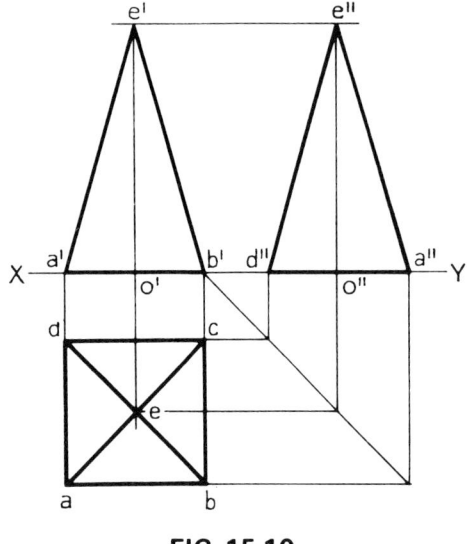

FIG. 15.10

FIG. 15.10 *To draw the plan and front and end elevations of a right square pyramid of base AB and axis EO with its base on the H.P., in front of and parallel to the V.P.*

The plan will be a square with its diagonals as the sloping edges of the pyramid, the front and end elevations both equal isosceles triangles. Draw the plan *abcd* on the H.P. making *ab* and *ad* both equal to *AB*. Draw projectors from the plan to the V.P. and the second V.P. and, between these and a horizontal line distance *EO* above the *XY*, draw the front elevation, *a'o'b'e'*, and the end elevation *d"o"a"e"*.

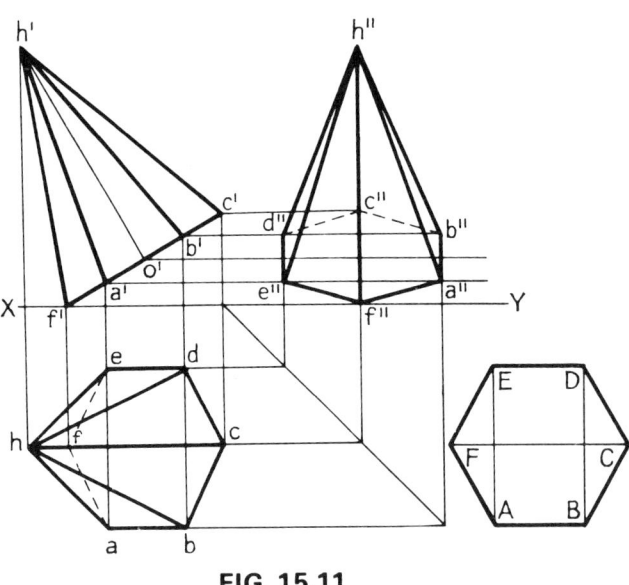

FIG. 15.11

FIG. 15.11 *To draw the plan and front and end elevations of a right hexagonal pyramid of base ABCDEF and axis OH. Point F is on the H.P., the base is inclined at 20° to the H.P., and edge ED is parallel to and in front of the V.P.*

As in Figs. 15.8 and 15.9, a separate additional view of the hexagonal base must be drawn first and then used to draw the front elevation. Because *f'c'* is equal to *FC*, the side *a'b'* and the axis *o'h'* are both true lengths. Draw the additional view, *ABCDEF*, and from it the front elevation, *f'a'o'b'c'h'*, on the V.P. Draw projectors from the front elevation to the H.P. and the second V.P. and draw the plan, *abcdefh*, between these projectors, making *ea* and *db* equal to *EA* and *DB* respectively. Draw projectors from the plan to the second V.P. and, through the points of intersection, the end elevation, *a"b"c"d"e"f"h"*.

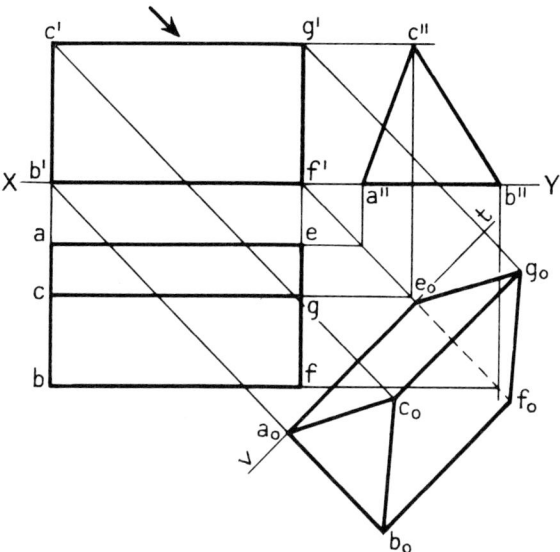

FIG. 15.12

FIG. 12. *To draw the plan, an auxiliary plan on an auxiliary perpendicular plane inclined at 45° to the H.P., front and end elevations of a right triangular prism of end ABC and length AE. Side ABFE is on the H.P. and edge AE is parallel to and in front of the V.P.*

Draw the end elevation and project the front elevation and plan from it. The auxiliary plan will be projected in a similar way to that used for laminae (Figs. 14.6 and 14.9), but, because a prism is greater than pencil-point thickness, the auxiliary plan will have an increased number of projectors and require more care in the transference of dimensions and joining points. Draw the v.t. on the H.P. and perpendicular projectors to it from the front elevation. Making g_0 and c_0 below the v.t. equal to eg or ac, a_0b_0 equal to ab, and e_0f_0 equal to ef. Complete the auxiliary plan $a_0b_0c_0e_0f_0g_0$ by joining the points on the projectors and lining in.

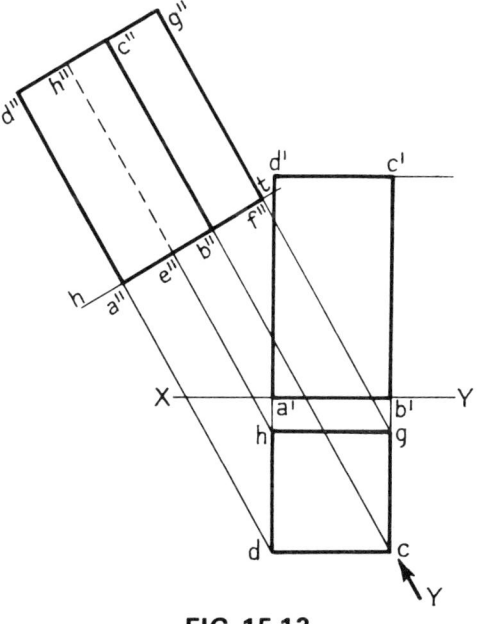

FIG. 15.13 *To draw the plan, front elevation, and auxiliary elevation, looking in the direction of arrow Y, of a right square prism, end AB and length AD. One end is on the H.P., parallel to and in front of the V.P.*

Draw the front elevation and plan as in Fig.15.6. The method of projection of the auxiliary elevation will be that used in Chapter 14 (laminae). Draw projectors from the plan and parallel to the direction of the arrow. Draw the h.t. perpendicular to the projectors, transfer $a'd'$ to the projector from d to give point d'' above a'' and draw $d''g''$ parallel to the h.t. to complete the auxiliary elevation $a''b''f''e''d''c''g''h''$.

FIG. 15.13

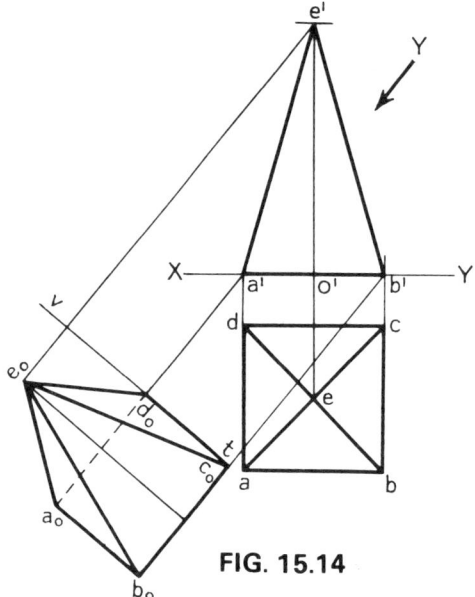

FIG. 15.14

FIG. 15.14 *To draw the plan, front elevation, and an auxiliary plan, looking in the direction of arrow Y, of a right square pyramid of base AB and axis EO. The base is on the H.P., in front of and parallel to the V.P.*

Draw the front elevation and plan. The auxiliary plan will be projected by the method used in Fig. 15.12. Therefore, from the front elevation, draw projectors parallel to the arrow *Y* and a v.t. on the H.P. perpendicular to these projectors. Make a_0d_0 and b_0c_0 equal to *ad* or *bc*, e_0 above the v.t. equal to *e* above *cd*, and join these points to complete the auxiliary plan $a_0b_0c_0d_0e_0$.

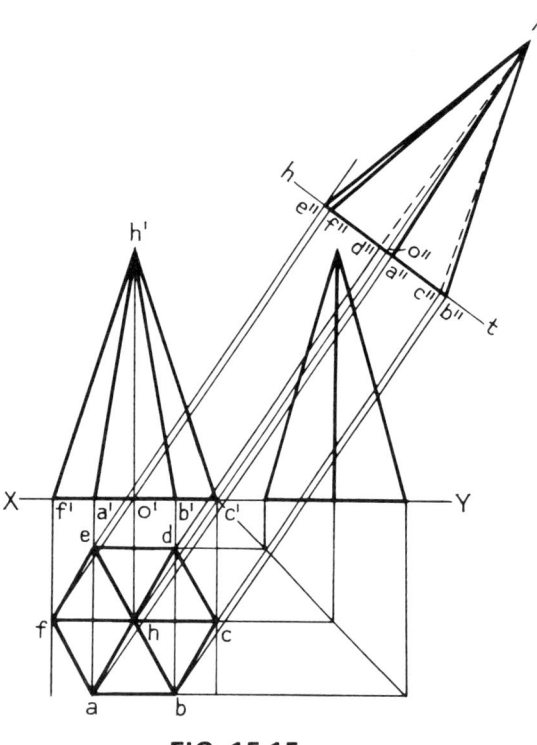

FIG. 15.15

FIG. 15.15 *To draw the plan, front elevation, end elevation, and an auxiliary elevation, on an auxiliary vertical plane inclined at 35° to the V.P., of a right hexagonal pyramid, base ABCDEF and axis HO. The base is on the H.P., one edge in front of and parallel to the V.P.*

Draw the front and end elevations on the V.P., and the plan on H.P., as in other examples. As in Fig. 14.8, draw the h.t. on the V.P. and inclined at 35°. From the plan draw projectors perpendicular to the h.t. and make $o''h''$ equal to $o'h'$. Join points on the h.t. to h'' to complete the auxiliary elevation.

EXERCISES

Draw the plan and front and end elevations of the following:

1. A cube of 45mm side with one surface on the H.P. and a second surface parallel to and 15mm in front of the V.P.

2. A cube of 52mm side with one edge on the H.P., one surface inclined at 30° to the H.P., a second surface parallel to and 15mm in front of the V.P.

3. A cube of 48mm side on the H.P., inclined at 30° to the V.P. and its nearest edge 12mm in front of the V.P.

4. A right square prism of 45mm side and length 75mm, one end on the H.P., 15mm in front of and parallel to the V.P.

5. A right regular pentagonal prism of 30mm side and 60mm axis. One rectangular surface is on the H.P., the axis parallel to and 68mm in front of the V.P.

6. An equilateral triangular right prism of 52mm side and 67mm axis, with one edge of an end on the H.P., a surface inclined at 30° to the H.P., parallel to and 15mm in front of the V.P.

7. A right regular hexagonal prism of 34mm side, 60mm axis, and with one surface on the H.P. The axis is inclined at 30° to the V.P. and the nearest end 45mm in front of the V.P.

8. A right regular octagonal prism of 26mm side and 75mm axis:

(i) One surface on the H.P., the axis inclined at 20° to and its nearest end 45mm in front of the V.P.

(ii) An edge of one end on the H.P., a surface inclined at 40° to the H.P., the axis parallel to and 45mm in front of the V.P.

9. A right square pyramid of 45mm side and 60mm axis, its base on the H.P., parallel to and 15mm in front of the V.P.

10. A right square pyramid of 48mm base and 68mm axis, its base inclined to the H.P. at 30°, one edge on the H.P., 15mm in front of and perpendicular to he V.P.

11. A right triangular pyramid of sides 38mm, 52mm, and 60mm, axis 68mm, its base on the H.P. the 60mm side inclined at 30° to the V.P. and its end 15mm in front of the V.P.

12. A right regular pentagonal pyramid of 36mm side and 68mm axis:

(i) Lying on one of its surfaces on the H.P., the base edge perpendicular to and its end 30mm in front of the V.P.

(ii) The base inclined at 20° to the H.P., one edge on the H.P., perpendicular to and one end 50mm in front of the V.P.

Draw a front elevation, plan, and auxiliary plan or elevation of the following:

13. A right square prism of 30mm side and 90mm axis, one surface on the H.P., 15mm in front of and parallel to the V.P.

(i) An auxiliary elevation on an auxiliary plane inclined at 30° to the V.P.

(ii) An auxiliary plan looking in the direction of an arrow inclined at 30° to the H.P.

14. A right regular heptagonal prism of 30mm side and 75mm axis, one surface on the H.P., the axis parallel to and 60mm in front of the V.P.

(i) An auxiliary elevation looking in the direction of an arrow inclined at 45° to the V.P.

(ii) A new plan on an auxiliary perpendicular plane inclined at 20° to the H.P.

15. A right regular hexagonal pyramid of 30mm side and 75mm axis, its base on the H.P., one side of the base parallel to and 15mm in front of the V.P.

(i) An auxiliary elevation on an auxiliary vertical plane inclined at 40° to the V.P.

(ii) An auxiliary plan looking in the direction of an arrow inclined at 20° to the H.P.

CHAPTER 16

PROJECTION OF SPHERES, CYLINDERS, AND CONES

Machines and buildings are not always entirely recti-linear and this chapter refers to common circular solids, all of which are often used as part of a Machine Drawing. The methods of projection are those of Chapter 15, with the addition of centre lines and ordinates, and are an extension of those used for circular laminae as in Chapter 14.

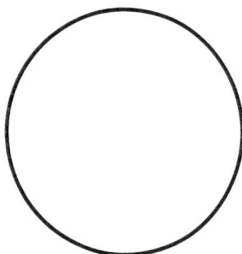

FIG. 16.1

Sphere, Right Cylinder, and Right Cone

FIG. 16.1 *A sphere is a solid of circular cross-section in all directions and generated by the revolution of a semicircle about its diameter.*

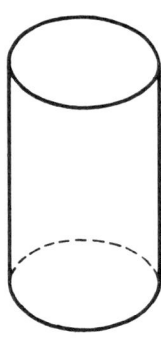

FIG. 16.2

FIG. 16.2 *A right cylinder is a type of prism but with equal circular ends. It is generated by the revolution of a rectangle about one long edge and its axis is at right angles to the ends and joins their centres.*

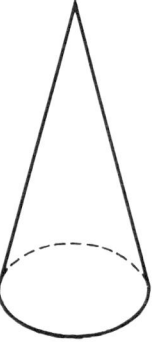

FIG. 16.3

FIG. 16.3 *A right cone is a type of pyramid but bounded by a cylinder base and a conical surface. The axis is perpendicular to and passes through the centre of the base and also the vertex (the highest point of the cone.)*

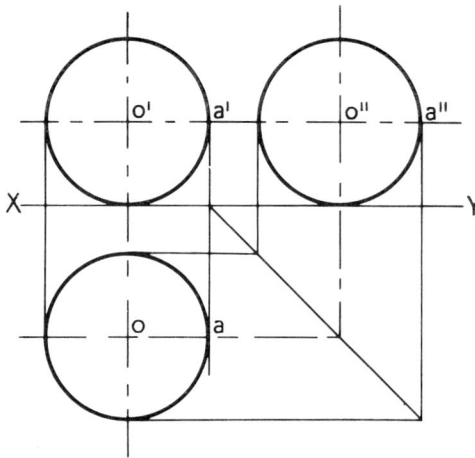

FIG. 16.4

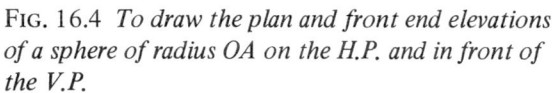

FIG. 16.4 *To draw the plan and front end elevations of a sphere of radius OA on the H.P. and in front of the V.P.*

Draw the horizontal centre line $o'o''$ on the V.P., distance OA above the XY, and the vertical centre line to mark the position of o'. Describe the circle $o'a$ and mark the position of o on the H.P. Draw a horizontal centre line through o and project it on to the second V.P. Describe circles at o and o''. From the definition of a sphere, all the views will be circles of equal radius.

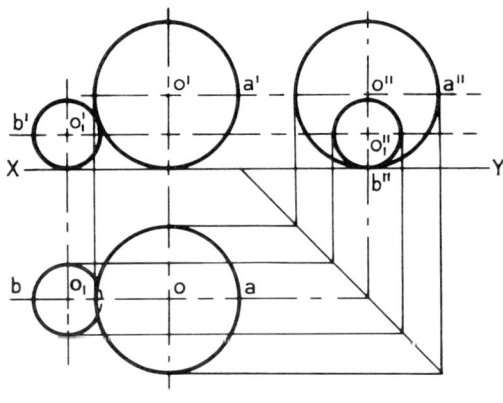

FIG. 16.5

FIG. 16.5 *To draw the plan and front and end elevations of two spheres, radii OA and OB, in contact, on the H.P. and in front of the V.P.*

As in Fig. 16.4, the views of each sphere will be circles of equal radius but part of the smaller sphere will be hidden by the larger and shown as hidden detail in the plan. Draw the centre lines on the V.P. to intersect at o' and o_1'—by the method given in Fig. 9.3—and describe the circles $o'a'$ and $o_1'b'$. Project the centre lines to the H.P., mark o and o_1, draw the horizontal centre line through o and project it to the second V.P. Project the horizontal centre lines from the front elevation to the second V.P. to give o'' and o_1'' and describe the circles oa, o_1b, $o''a''$, and $o_1''b''$ showing part of o_1b as hidden detail.

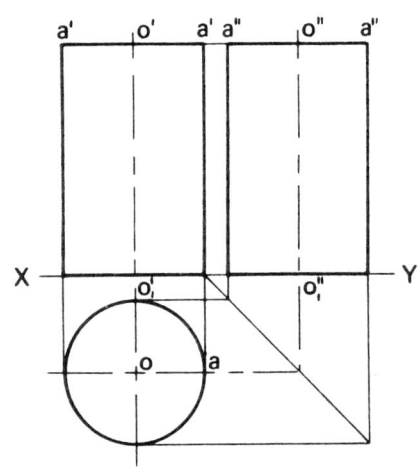

FIG. 16.6

FIG. 16.6 *To draw the plan and front and end elevations of a right cylinder of radius OA and axis OO_1. The cylinder is in front of the V.P. with one end on the H.P.*

The elevations will be equal rectangles and the plan a circle. Draw the centre lines on the H.P., to intersect at o, and project them to the V.P. and the second V.P. Describe the circle oa and from it draw protectors to the V.P. and the second V.P. Draw the horizontal line $a'o'a'$ and project it to $a''o''a''$ to complete the elevations.

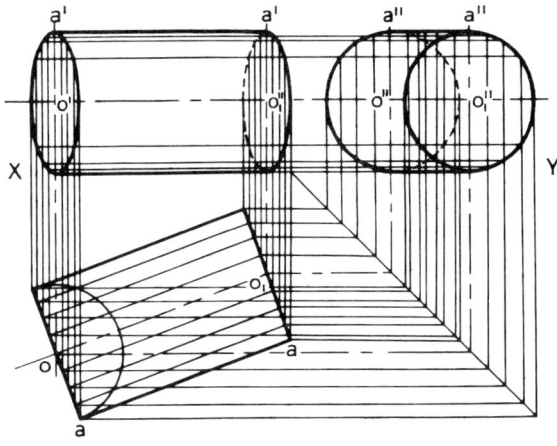

FIG. 16.7

FIG. 16.7 *To draw the plan and front and end elevations of a right cylinder of radius OA and axis OO$_1$, lying on the H.P., in front of the V.P., its axis inclined at 20° to the V.P. and parallel to the H.P.*

Each elevation will consist of a shortned side and elliptical ends, and a true view of one end (in this example a semicircle on the plan) will be necessary in order to project the elevations. The method is similar to those of Figs. 14.4 and 15.9. Draw the centre line and plan on the H.P., making oo_1 equal to OO_1 and oa equal to OA, and describe a semicircle on oa. Draw equally spaced ordinates on oa and produce them the length of the plan and parallel to oo_1. From the ends of the plan project the centre lines and ordinates to the V.P. and the second V.P. and draw the centre line $o'o_1''$ distance OA above the XY. From each side of the centre line $o'o_1''$ and on the vertically projected ordinates, step off the ordinate lengths of the semi-circle oa, and through these points draw the curves $o'a'$, $o'a''$, $o''a''$, and $o_1''a''$. Complete the elevations by joining $a'a'$ and $a''a''$ and indicating hidden detail.

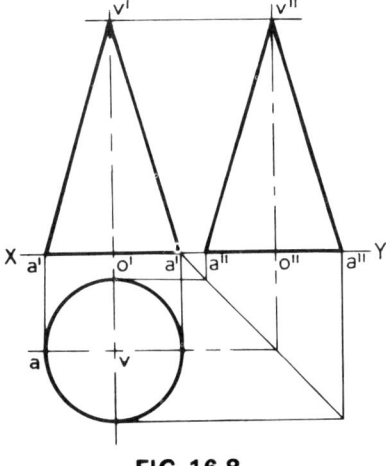

FIG. 16.8

FIG. 16.8 *To draw the plan and front and end elevations of a right cone of radius OA and axis OV, its base on the H.P. and in front of the V.P.*

The elevations will be equal triangles and the plan a circle. Draw the centre lines on the H.P., intersecting at v, and project them to the V.P. and the second V.P. With v as centre, describe a circle of radius OA and draw projectors from this to the V.P. and the second V.P. On the V.P. draw $v'v''$ parallel to the XY and distance OV above it; join $v'a'$ and $v''a''$ to complete the elevations.

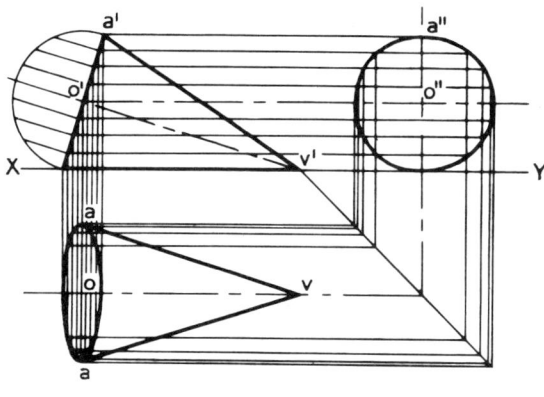

FIG. 16.9

FIG. 16.9 *To draw the plan and front and end elevations of a right cone of radius OA and axis OV, lying on its side on the H.P., in front of and its axis parallel to the V.P.*

As in Fig.16.7, an additional view will have to be drawn before the plan and end elevation can be projected. Draw the front elevation, making $o'v'$ equal to the axis OV and $o'a'$ equal to the radius OA, and on $o'a'$ equal to the radius OA, and on $o'a'$ describe a semicircle to represent the additional view of the base. Draw equally spaced ordinates on the semicircles and project these, the ends of the centre line $o'v'$ and a' to the H.P. and the second V.P. Draw the centre line ov on the H.P., and on the vertically projected ordinates and each side of ov step off the ordinate lengths from the semicircle, draw a curve through the points, and complete the plan by joining av. Project the points on the plan to the second V.P. and through the intersection of these projectors with those from the front elevation, draw the end elevation $o''a''v''$.

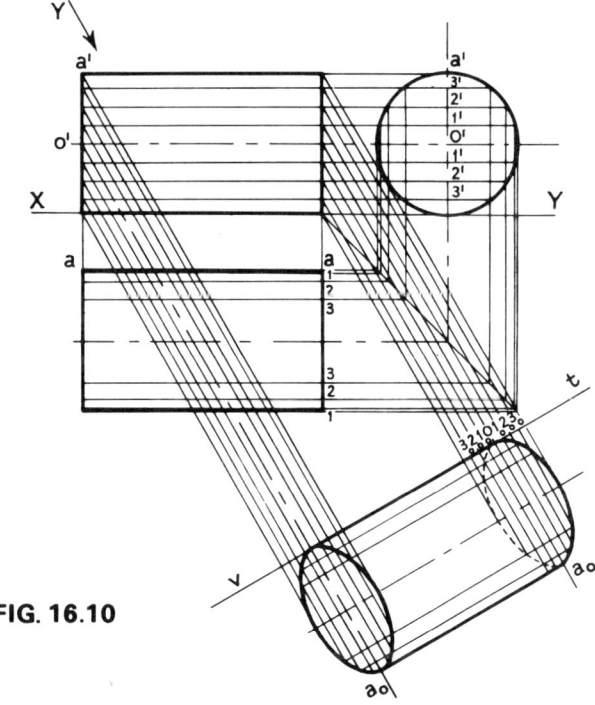

FIG. 16.10

FIG. 16.10 *To draw the front and end elevations, plan and an auxiliary plan, looking in the direction of arrow Y, of a right cylinder of radius OA and axis OO_1. The cylinder is on the H.P., in front of and parallel to the V.P.*

Draw the front and end elevations on the H.P., as in other examples. Draw horizontal ordinates on the end elevation and project these to the plan and front elevation. The auxiliary plan will be projected from the front elevation, as in Chapters 14 and 15. Draw the v.t. perpendicular to the direction of the arrow Y and projectors from the front elevation parallel to arrow Y, perpendicular to and below the v.t. Taking distances from the plan, make the ordinates $3_0, 2_0, 1_0, 1_0, 2_0,$ and 3_0 above the v.t. equal to $a3, a2, a1, a1, a2,$ and $a3$ and similarly the distance above the v.t. of the centre line and a_0a_0. Draw a smooth curve through the points obtained to complete the auxiliary plan.

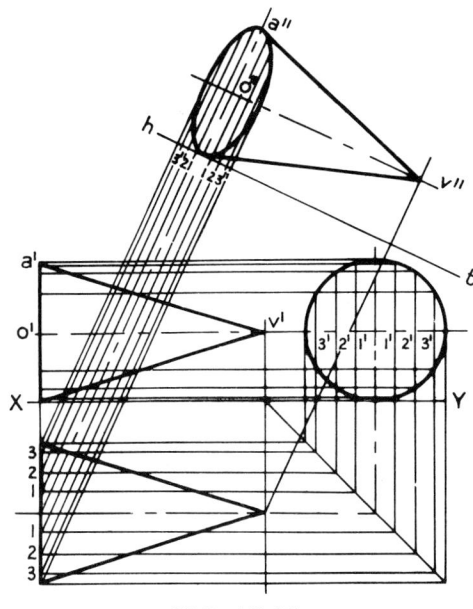

FIG. 16.11

FIG. 16.11 *To draw a plan, front and end elevations, and an auxiliary elevation, on an auxiliary vertical plane inclined at 25° to the V.P., of a right cone of radius OA and axis OV. Its base is perpendicular to the H.P., one point in contact with the H.P., in front of the V.P., and its axis is parallel to the H.P. and the V.P.*

Draw the front and end elevations on the V.P. and the plan on the H.P. Draw ordinates vertically on the end elevation and project them to the plan and front elevation. As in previous examples, the auxiliary elevation will be projected from the plan, therefore draw the h.t. of the auxiliary plane at an angle of 25° and projectors from the plan perpendicular to and above the h.t. Mark off the lengths of the ordinates in elevation on the projected ordinates 3″, 2″, 1″, 1″, 2″, and 3″ and the distances of *v″* and *a″* above the h.t. equal to *v′* and *a′* above the *XY*. Join these points to complete the auxiliary elevation.

EXERCISES

Draw the plan and front and end elevations of the following:

1. A sphere of 36mm radius on the H.P. and its centre 52mm in front of the V.P.

2. Two spheres of radii 33mm and 22mm, both in contact on the H.P. and on a common centre line parallel to and 45mm in front of the V.P.

3. Three spheres of radii 52mm, 18mm, and 15mm, all three are on the H.P. in the order of 18mm, 52mm, and 15mm, the 52mm sphere touching the 18mm and 15mm spheres, their centre lines parallel to and 68mm in front of the V.P.

4. A right cylinder of 22mm radius and 60mm axis, a point on one end touching the H.P., the end inclined at 30° to the H.P., the axis parallel to and 36mm in front of the V.P.

5. A right cylinder of 30mm radius and 82mm axis:
(i) On the H.P. with the axis inclined at 30° to the V.P. and the nearest end 40mm in front of the V.P.

(ii) The axis inclined at 20° to the H.P., parallel to and 45mm in front of the V.P. and one point of one end on the H.P.

6. A right cone of 26mm radius and 60mm axis, with its base on the H.P. and its axis parallel to and 35mm in front of the V.P.

7. A right cone of 22mm radius and 52mm axis:
(i) On its side on the H.P., its axis parallel to and 35mm in front of the V.P.
(ii) One point of the base on the H.P., the base inclined at 15° to the H.P. and its axis parallel to and 35mm in front of the V.P.

8. A right cone of 26mm radius and 60mm axis, its axis parallel to the H.P. and the V.P., 26mm above the H.P. and 40mm in front of the V.P., with:
(i) An auxiliary elevation on an auxiliary vertical plane inclined at 30° to the V.P.
(ii) An auxiliary plan looking in the direction of an arrow inclined at 40° to the H.P.

CHAPTER 17

SECTIONS OF RECTILINEAR SOLIDS BY VERTICAL, HORIZONTAL, AND INCLINED PLANES

In machine parts, internal-combustion engines, electric motors, buildings, etc., elevations and plans do not show clearly the insides and therefore do not give sufficient information to enable manu-facture or construction to take place. This difficulty is overcome by introducing imaginary cutting planes and then drawing what is seen when the cut-off part has been removed. These are called Sectional Views. The position or trace of the plane is indicated by a Cutting Plane Line and the cut part by cross-hatching at 45°. In cases where the section does not give a true shape, auxiliary sections can be projected by the same method as that used for auxiliary elevations and plans.

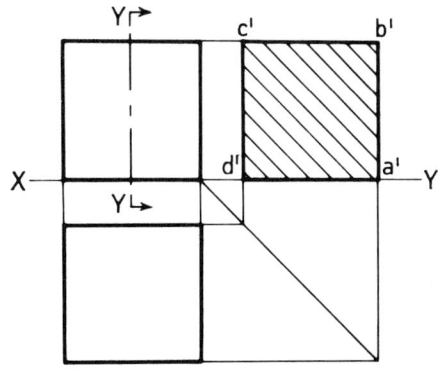

FIG. 17.1

FIG. 17.1 *To draw the plan and front and sectional elevations of a cube of side AB sectioned on the line YY.*

The line *YY* is the standard form of section line and the arrows indicate the direction in which the remainder of the cube should be viewed—the line *YY* is the vertical trace of the cutting plane. The arrows are not essential in this example, because they point in the same direction as that used in first angle projection, but are often included to indicate the correct direction. The sectioned view is of the same dimensions as the end elevation. Therefore, by the method of Fig. 15.4, draw the plan and front and end elevations. To complete the section *a'b'c'd'*, draw equally spaced lines at 45° across the end elevation—this is cross-hatching and is the general method of indicating a cut surface.

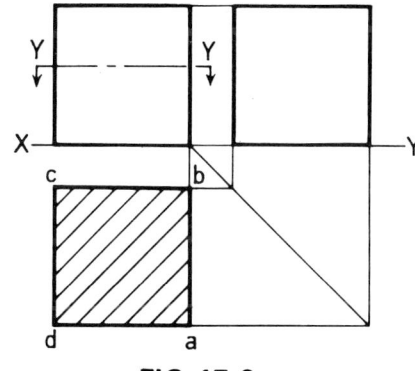

FIG. 17.2

FIG. 17.2 *To draw the sectional plan and front and end elevations of a cube of side AB cut by a horizontal plane on YY.*

Because the cutting plane is horizontal, the sectioned part will be a plan equal in size and shape to the plan of the cube above the cutting plane. By the method used in Fig. 17.1, draw and project the sectioned plan and front and end elevations and complete the sectioned view, *abcd*, by cross-hatching at 45°.

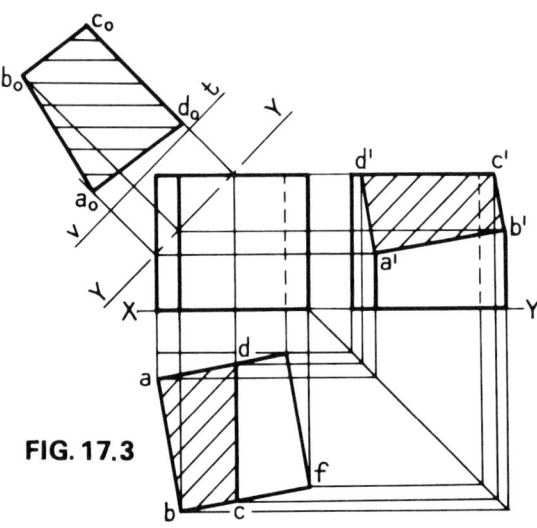

FIG. 17.3

FIG. 17.3 *To draw the sectioned views and the true sectional shape of a cube cut by a perpendicular plane inclined to the H.P. The cube is on the H.P., inclined to and in front of the V.P.*

In this example only part of the plan and end elevation will be cut by the plane, therefore draw the plan and both elevations. Draw the cutting plane line *YY* (i.e. the v.t. of the plane) and draw projectors, horizontally and vertically, from the points where *YY* cuts the edges of the cube. Complete the section *abcd* in plan, draw projectors from *abcd* to the end elevation and join the points of intersection to make the section $a'b'c'd'$. Neither section gives the true shape, but it can be found by drawing an auxiliary plan on a parallel perpendicular plane to *YY*. Using the methods of Chapters 14, 15, and 16, draw a parallel v.t. to *YY* and perpendicular projectors from *YY*. Transfer the perpendicular distances of *a*, *b*, *c*, and *d* from the plan to give a_0, b_0, c_0, and d_0 above the new v.t. Join a_0 b_0 c_0 d_0 and cross-hatch the three views at $45°$ to their planes of projection.

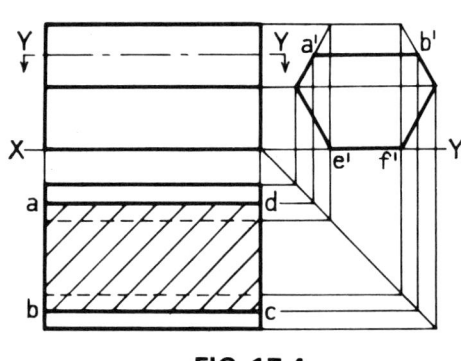

FIG. 17.4

FIG. 17.4 *To draw the sectional views of a right hexagonal prism of side EF and length AD, cut by a horizontal plane YY on the H.P., parallel to and in front of the V.P.*

Draw the end elevation and from it project the plan and front elevation. Draw the cutting plane line *YY* and project it to the end elevation to cut off the top part on $a'b'$. From $a'b'$ draw projectors to the plan to give the section *abcd* and cross-hatch to complete the view.

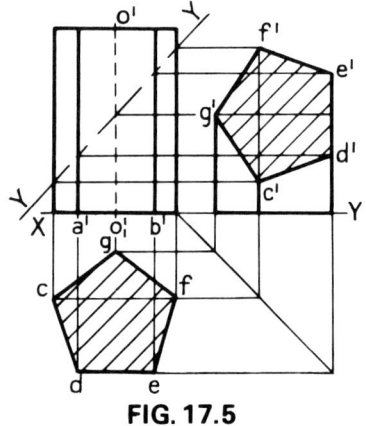

FIG. 17.5

FIG. 17.5 *To draw the sections of a right pentagonal prism of side AB and axis OO_1, with one end on the H.P., in front of and parallel to the V.P. and cut by a perpendicular plane YY inclined to the H.P.*

Neither of these sectional views is a true shape of the section, but the sectional plan is the same as the plan because the prism is cut between the ends. Draw the plan, from it project the front and end elevations and draw the cutting plane line *YY*. Draw projectors from the points where *YY* cuts the edges of the prism to the end elevation and join $c'd'e'f'g'$ to complete the end elevation of the prism below the cutting plane. Cross-hatch $c'd'e'f'g'$ and *cdefg*.

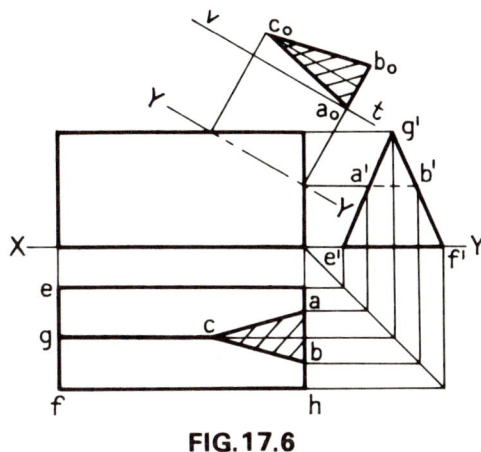

FIG. 17.6

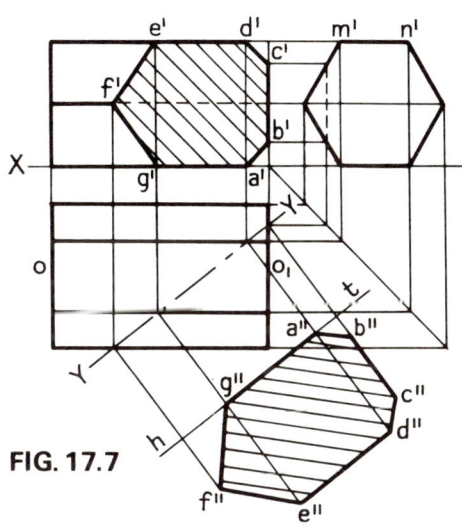

FIG. 17.7

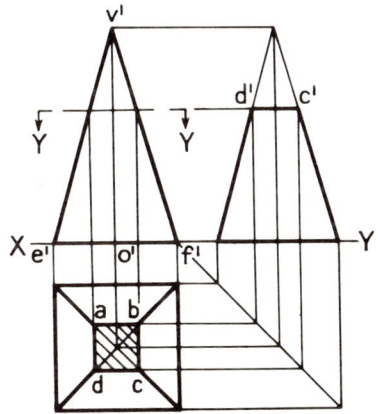

FIG. 17.8

FIG. 17.6 *To draw the sectional views and a true section of a right triangular prism of end EFG and length FH, with side EFH on the H.P., parallel to and in front of the V.P., and cut by a perpendicular plane YY inclined to the H.P.*

Draw the end elevation and from it project the front elevation and plan. Draw the cutting plane line *YY* and projectors to the end elevation and plan from the points where *YY* cuts the edges of the prism. The projectors give $a'b'$ on the end elevation—this is hidden detail because the prism is cut on the end nearest the end elevation—and *abc* on the plan. The true shape, as in Fig. 17.3 is an auxiliary plan of the section and will be drawn on a parallel v.t. to *YY*. Draw the v.t., perpendicular projectors from *YY*, and transfer the distances of *b* and *c* above the horizontal projector passing through *a* to give the points a_0, b_0, and c_0. Join these points and cross-hatch the sectional views.

FIG. 17.7 *To draw the sectional views and a true section of a right hexagonal prism of end MN and axis OO_1, with one side on the H.P., the prism in front of the V.P. and the axis parallel to the V.P. The prism is cut by a vertical cutting plane YY inclined to the V.P.*

Draw the end elevation and project the plan and front elevation. Since the cutting plane is inclined to the vertical, its trace (h.t.) is on the plan, therefore draw *YY* and projectors from it to the elevations. Show the section by hidden detail on the end elevation, project from it to the front elevation and join $a'b'c'd'e'f'g'$ to complete the sectional view. The true shape will be an auxiliary elevation and drawn by the method used in Figs. 17.3 and 17.6. Draw a parallel h.t. to *YY*, perpendicular projectors from *YY* to h.t., and transfer the perpendicular distances of b', c', d', e', and f' above a' and g' to the projectors below the h.t. Join a'', b'', c'', d'', e'', f'', and g'' to complete the true shape and cross-hatch.

FIG. 17.8 *To draw the sections of a right square pyramid of base EF and axis OV. The base is on the H.P., in front of and parallel to the V.P., and cut by a horizontal plane YY.*

Draw the plan and both elevations as in Fig. 15.10. Draw the cutting plane line *YY*, projectors from it to the plan and end elevation and then from the end elevation to the plan. Join *a*, *b*, *c*, and *d*, remove the part of the pyramid above $c'd'$, and cross-hatch the sectional part of the plan.

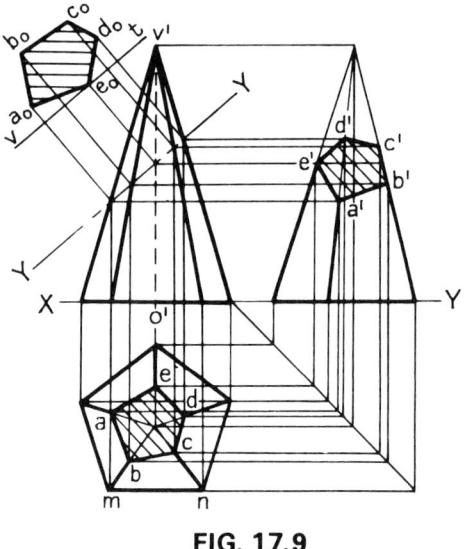

FIG. 17.9

FIG. 17.9 *To draw the sectional views and a true section of a right pentagonal pyramid of base MN and axis OV. The base is on the H.P., in front of and one edge parallel to the V.P., and cut by a perpendicular plane YY inclined to the H.P.*

Draw the plan and project the front and end elevations. Draw YY and, from the points where it cuts the edges of the pyramid, projectors to the end elevation and plan. Join a', b', c', d', and e', draw projectors from them to the plan and join the points of intersection, a, b, c, d, and e. The true shape is an auxiliary plan and, as in other examples, draw a parallel v.t. to YY and projectors from YY to the v.t. Transfer the distances of a, b, c, and d, below the horizontal projector passing through e, to above the v.t. and join a_0, b_0, c_0, d_0, and e_0. Cross-hatch the sections.

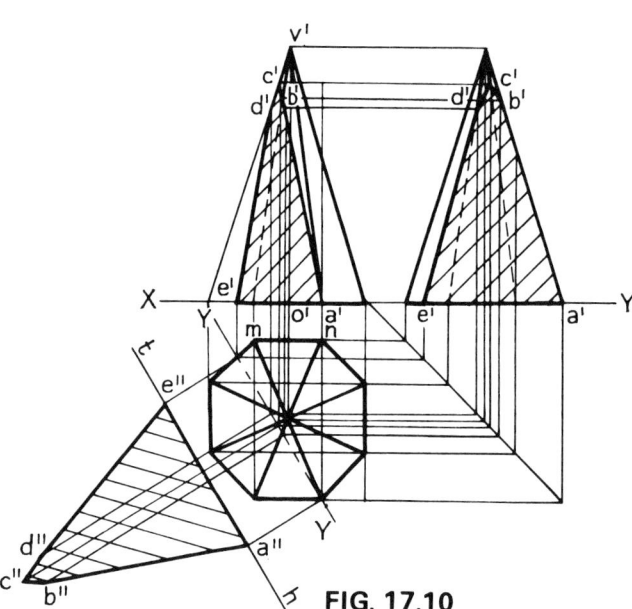

FIG. 17.10

FIG. 17.10 *To draw the sectional views and a true section of a right octagonal pyramid of base MN and axis OV, the base on the H.P., in front of and one edge parallel to the V.P. and cut by a vertical plane inclined to the V.P.*

Draw the plan and project the front and end elevations. Draw the cutting plane line YY and, from the points where it cuts the edges of the pyramid, projectors to the two elevations. Draw projectors from the front elevation to the end elevation, remove the part pyramid in front of YY on the front elevation and join a', b', c', d', and e' in both elevations. The true shape will be an auxiliary elevation projected on to a parallel h.t. to YY. Draw a parallel h.t. to YY, perpendicular projectors from YY, and transfer the distances of b', c', and d' above $a'e'$ to the projectors passing through h.t. to give points b'', c'', and d''. Join a'', b'', c'', d'', and e'' and cross-hatch the sectional views.

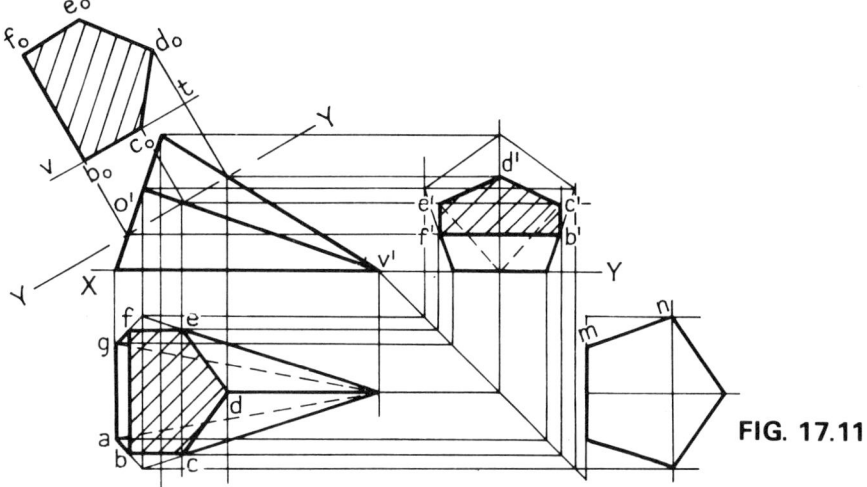

FIG. 17.11

FIG. 17.11 *To draw the sectional views and a true section of a right pentagonal pyramid of base MN and axis OV, with one side on the H.P., the pyramid in front of and the axis parallel to the V.P. and cut by a perpendicular plane inclined to the H.P.*

As in Chapters 14 and 15, an additional view of the base must be drawn in order to project the other views. Draw the additional view, project the plan, front and end elevations and draw the cutting plane line YY. From the points where YY cuts the edges of the pyramid, draw projectors to the plan and end elevation and join the points of intersection on the plan to give the shape $bcdef$. Project $bcdef$ to the end elevation and through the intersections draw the shape $b'c'd'e'f'$. The true shape of the section is an auxiliary plan, therefore draw a v.t. parallel to YY and perpendicular projectors through v.t. Transfer the distances of b, c, d, and f, above the projector passing through e, to the projectors above the v.t. Join b_0, c_0, d_0, e_0, and f_0 and cross-hatch the views.

Dihedral angle

One of the very important sections of a solid, which has adjoining sloping faces, is that containing the dihedral angle, i.e. the true angle between any two adjoining faces (oblique planes).

In Fig. 17.12 the front elevation and plan of a right square pyramid, of base $ABCD$ and axis OV, are given. By the method given in Fig. 12.10, the true length of the edge AV can be found. The horizontal trace of the vertical plane containing AV is av and vV is the vertical height of V above O (equal to $o'v'$) and on bd which is at right angles to ac by construction. Join AV, then the right-angled triangle AvV is a true vertical section on the edge AV, giving the true length and the true angles of inclination of AV to the base and axis of the pyramid. This method of drawing a true section is only applied when a true length and true angles are required and therefore, in practice, is never cross-hatched.

In Fig. 17.12 the true section gives a true length and true angles of inclination, but not the true angles between two adjoining faces. Since a dihedral angle is found by cutting across AV with a section plane whose traces are perpendicular to the edge AV and plan av, then the drawing of the section of Fig. 17.12 is the first step in finding a dihedral angle.

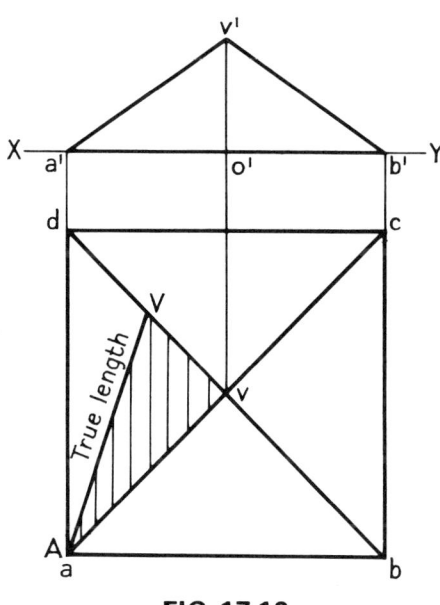

FIG. 17.12

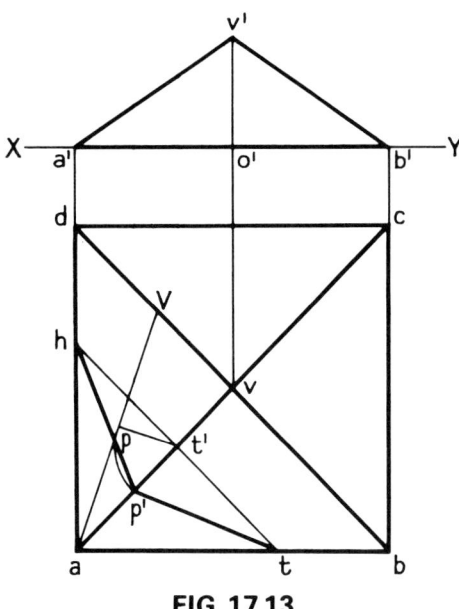

FIG. 17.13

FIG. 17.13 *To find the dihedral angle of a right square pyramid of base ABCD and axis OV.*

Draw the front elevation, $a'o'b'v'$, and the plan *abcdv,* of the pyramid. Since *avd* is a right angle, mark off *vV* equal to $o'v'$ and join *Va* to complete a true vertical section as in Fig. 17.12. Draw the horizontal trace *ht* of the cutting plane, on which will be the dihedral angle, perpendicularly across *av* and to cut it at any convenient point, t'. The vertical trace of this auxiliary plane will be perpendicular to *aV*, therefore draw pt' perpendicular to *aV*. Since pt' is the v.t. of the auxiliary plane within the pyramid, then rabat pt' to p' on *av* (because the plan of point *p* on edge *aV* lies on *av*), join p' to *h* and *t* to complete the section formed by the auxiliary plane. The angle $hp't$ is the dihedral angle or true angle between the faces *ABV* and *ADV* of the pyramid.

The dihedral angle is required in roof construction (e.g. the angle between the top edges of a hip) and the production of *L*-section metal castings or strip which is to be placed where it will slope in two directions (e.g. the sloping legs of a machine stand).

EXERCISES

With two of the views sectioned, draw a plan and front and end elevations of the following:

1. A cube of 60mm edge, on the H.P., 15mm in front of and parallel to the V.P.:
(i) Cut by a vertical plane, perpendicular to the V.P. and 30mm from one side.
(ii) Cut by a horizontal plane 35mm above the H.P.
(iii) Cut by a perpendicular plane, inclined at $45°$ to the H.P. and cutting the elevations 30mm above the H.P. Add a true shape of the section.

2. A hexagonal prism of 26mm edge and 75mm axis, one side on the H.P., 30mm in front of and parallel to the V.P. and cut by:
(i) A horizontal plane 35mm above the H.P.
(ii) A perpendicular plane, inclined at $40°$ to the H.P. and cutting the top 45mm from one end.
(iii) A vertical plane inclined at $60°$ to the V.P. and 35mm from one end. Add the true shape of the section.

3. A right pentagonal prism of 30mm edge and 68mm axis, with one end on the H.P., one side parallel to the V.P., the axis 60mm in front of the V.P., and cut by:
(i) A perpendicular plane inclined at $20°$ to the H.P. and cutting the axis of the prism 45mm above the H.P. Add a true shape of the section.
(ii) A vertical plane inclined $20°$ to the V.P. and passing through the axis. Add a true shape of the section.

4. A right square pyramid of 82mm axis and 63mm base, with one triangular surface on the H.P., its axis parallel to and 45mm in front of the V.P.
(i) A horizontal plane 33mm above the H.P.
(ii) A vertical plane, inclined at $40°$ to the V.P. and passing through the base edge on the H.P., 56mm in front of the V.P.

5. A right hexagonal pyramid of 68mm axis and 36mm edge, with the base on the H.P., the nearest edge parallel to and 15mm in front of the V.P. and cut by:
(i) A perpendicular plane inclined at $30°$ to the H.P. and passing through the mid-point of the axis.
(ii) A vertical plane parallel to the V.P. and 68mm in front of the V.P.

6. A right pentagonal pyramid of 105mm axis and base *ABCDE*, where *AB* = 45mm *BC* = 48mm, *AC* = 67mm, *CD* = 52mm, *BD* = 75mm, *AE* = 37mm, and *DE* = 42mm. The vertex is perpendicularly above the intersection of *AC* and *BD*. The base is on the H.P., edge *AB* parallel to and 15mm in front of the V.P. and cut by:

(i) A perpendicular plane inclined at 30° to the H.P. and cutting the axis 52mm above the H.P. Add the true shape of the section.

(ii) A vertical plane cutting the base through *E* and *C*. Add the true shape of the section.

7. Draw the dihedral angle of a right square pyramid of 90mm base and 60mm axis. Show the true length and angles of inclination of one edge.

CHAPTER 18

SECTIONS OF CONES AND CYLINDERS BY PERPENDICULAR PLANES

Because Machine Drawings are very rarely entirely rectilinear, it is essential to know how to section the common circular solids. The methods are almost the same as those used in Chapter 17, but the cutting plane passes through curved surfaces, instead of cutting edges, and therefore points of projection have to be given. The approach is to regard all projectors as cutting planes, then if projectors (imaginary planes) are placed in convenient positions the section cutting plane will pass through two points on each imaginary plane, so giving a series of points on the curve of the section.

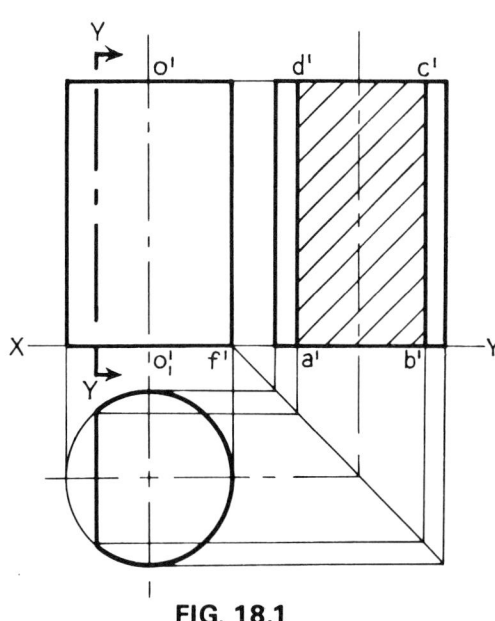

FIG. 18.1

FIG. 18.1 *To draw the section of a right cylinder of axis OO_1 and radius OF, one end on the H.P., in front of the V.P. and cut by a vertical plane, YY, perpendicular to the V.P.*

In this example the section plane gives a true sectional shape in the end elevation, which is a rectangle and does not require the use of imaginary cutting planes. Draw the plan, both elevations, and the cutting plane line *YY*. Project *YY* to the plan, from the points of intersection on the plan, draw projectors to the end elevation to give section $a'b'c'd'$ and cross-hatch the view.

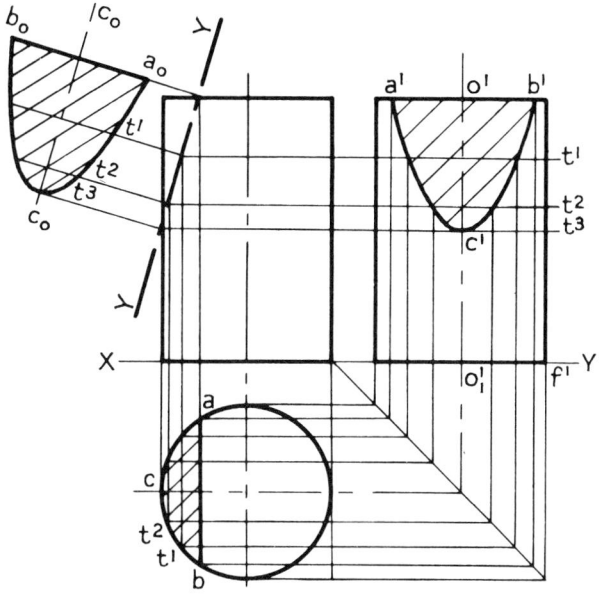

FIG. 18.2

FIG. 18.2 *To draw the sectional views and a true section of a right cylinder of axis OO_1 and radius OF. One end is on the H.P., in front of the V.P. and cut by a perpendicular plane inclined to the H.P.*

To draw the section produced by this cutting plane, a series of points will have to be found at different levels on the cylinder. There will be two at each level where the cutting plane (YY) and the circumference of the cylinder intersect. Draw the plan, front and end elevations of the cylinder, and the section line YY. At convenient levels (at points where the curves have small radii) and where YY cuts the surface, draw the v.t.s, t^1, t^2, and t^3, of imaginary horizontal planes. Draw projectors from the intersection of the planes with YY to the plan and end elevation and complete the section in plan, *abc*. From the points of intersection of the projectors on the plan, draw projectors to the end elevation and through points on t^1, t^2, t^3 a smooth curve to give section $a'b'c'$. The true shape is an auxiliary plan, but because the shape will be a symmetrical curve, draw a centre line parallel to YY in place of a v.t. Draw perpendicular projectors from YY, and on these, and about the centre line, step off the lengths of t^1, t^2, and b to the centre line on the plan. Join a_0b_0, draw a smooth curve through the other points and cross-hatch the sections.

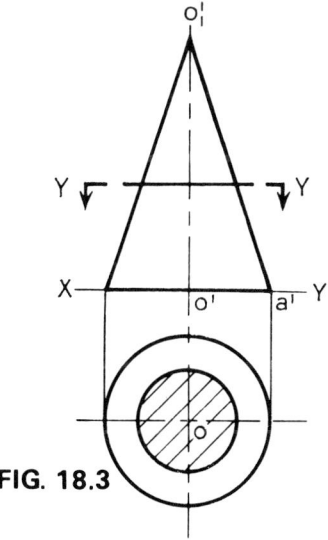

FIG. 18.3

FIG. 18.3 *To draw the section of a right cone of radius OA and axis OO_1, the base on the H.P., in front of the V.P. and cut by a horizontal cutting plane YY.*

The method in this example is similar to that of Fig. 17.8. Draw the centre lines and plan and project the front elevation. Draw the cutting plane line YY and projectors to the plan from the points where YY cuts the surface of the cone. With centre *o*, draw a circle on the plan equal in diameter to the distance between the projectors and cross-hatch this sectional view. By reference to Fig. 16.8, it will be seen that an end elevation is not necessary to provide details of the size and shape of a cone and in this example it has been omitted.

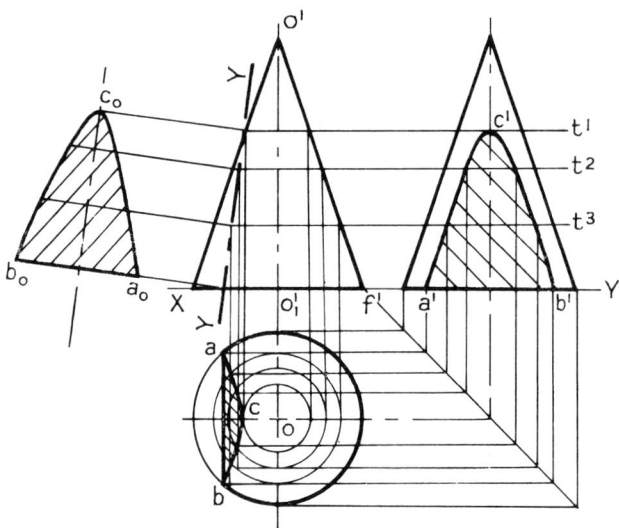

FIG. 18.4

FIG. 18.4 *Hyperbola. To draw the sectional views and a true section of a right cone of radius OF and axis* OO_1, *on the H.P. in front of the V.P., and cut by a perpendicular plane of greater angle of inclination to the H.P. than that of the side of the cone.*

Figs. 18.4, 18.5, and 18.6 all require imaginary cutting planes, the section of each one drawn on the plan and, because the true sections are symmetrical curves, centre lines in place of v.t.s.

Draw the plan and project the front and end elevations. Draw YY, the v.t.s of the imaginary horizontal cutting planes, t^1, t^2, and t^3, and projectors from their intersections with YY to the plan. On the plan draw three circles, each equal to the diameter of the cone on t^1, t^2, and t^3, to represent the sections of the horizontal planes and, through the intersections of the projectors from YY with the circles, a smooth curve to give the section abc. From the points on abc draw projectors to the end elevation and, through the intersections on t^1, t^2, and t^3, the section $a'b'c'$. As in Fig. 18.2, draw a centre line parallel to YY, perpendicular projectors to the centre line, and about it, step off the lengths on the plan from the centre line to the curve. Complete the true shape by joining a_0b_0 and drawing a smooth curve through the points on t^1, t^2, and t^3 projected. Cross-hatch the sectional views. The true shape, as an open curve, is a Hyperbola and is of considerable importance in many ways. In industry, for example, it is used as a means of directing and shaping a beam of light, such as a motor-car headlamp.

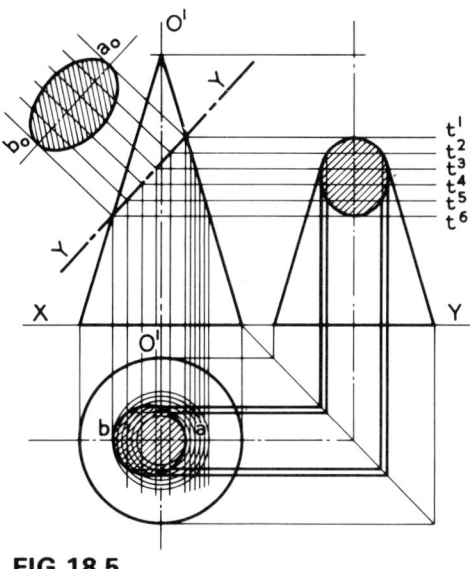

FIG. 18.5

FIG. 18.5 *Ellipse. To draw the sectional views and a true section of a right cone of radius OF and axis OO_1, its base on the H.P., in front of the V.P. and cut by a perpendicular plane inclined to the H.P.*

Draw the plan, project the front and end elevations, and draw YY. Draw the v.t.s of the imaginary horizontal cutting planes, two where YY cuts the side of the cone, t^1 and t^6, and four intermediate, t^2, t^3, t^4, and t^5, because of the small curves. Draw projectors from YY to the plan and six circles on the plan to represent the sections of the horizontal planes. Through the intersections of the circles with the projectors from YY, draw a smooth curve which gives the sectional plan, *ab*. From the points on *ab*, draw projectors to the end elevation, and through the intersections on t^1, t^2, t^3, t^4, t^5, and t^6, draw the section $a'b'$. Draw a centre line parallel to YY, perpendicular projectors from YY, and about the centre line, step off the lengths on the plan from the centre line to the curve. Join the points on the t^1, t^2, t^3, t^4, t^5, and t^6 to give the true shape a_0b_0 and cross-hatch the sectional views. This true shape is an Ellipse, now derived as the section of a cone instead of the locus of a point.

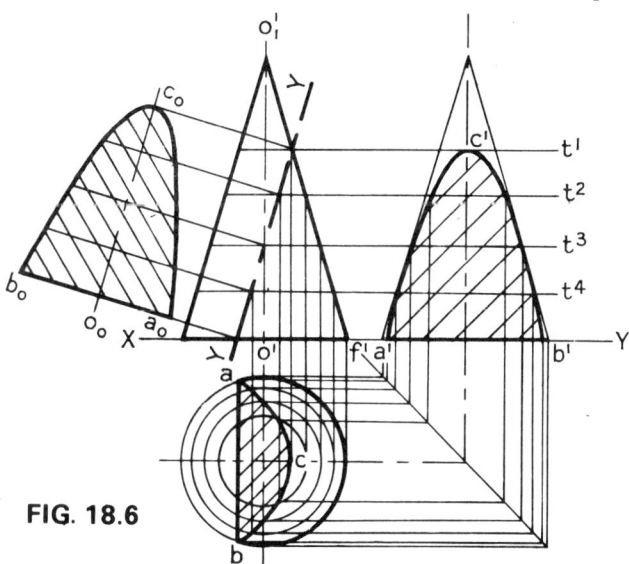

FIG. 18.6

FIG. 18.6 *Parabola. To draw the sectional views and true section of a right cone of radius OF and axis OO_1, its base on the H.P., in front of the V.P. and cut by a perpendicular plane YY parallel to the slant surface of the cone.*

Draw the plan, project the two elevations, and draw the cutting plane line YY and the v.t.s of the imaginary horizontal cutting planes, t^1, t^2, t^3, and t^4. Draw projectors, from the intersections of the v.t.s with YY, to the plan and circles on the plan to represent the sections of the horizontal planes. Through the intersections of the projectors from YY with the circles, draw the curve *abc*, to complete the sectional plan and then projectors to the end elevation. Join the points on t^1, t^2, t^3, t^4, a', and b' to complete the sectional end elevation. Draw a centre line parallel to YY, perpendicular projectors from YY to the centre line, and step off the lengths from the centre line to the curve in the plan on to the projectors of t^2, t^3, t^4, and a_0b_0, about the centre line c_0o_0. Join the points $a_0b_0c_0$ with a smooth curve to complete the true shape. Cross-hatch the sectional views. This true shape, as an open curve, is a Parabola. It is the path of a projectile and is used in industry in the manufacture of parabolic lenses, etc.

EXERCISES

Draw the sectional views and a true section of the following:

1. A right cylinder of 45mm diameter and 60mm axis:
(i) One end is on the H.P., the centre line (or axis) 38mm in front of the V.P., is cut by a perpendicular plane inclined at 60° to the H.P. and the plane passes through the centre of the top end.
(ii) On the H.P., the axis parallel to the H.P. and the V.P., 36mm in front of the V.P. and cut by a perpendicular plane inclined at 30° to the H.P.,

15mm above the H.P. at one end of the cylinder.
2. A cone of 90mm diameter, axis 105mm, with base on the H.P., centre 60mm in front of the V.P. and cut by:
(i) A perpendicular plane parallel to the slant side and passing through a point 18mm from the circumference of the base.
(ii) A perpendicular plane, inclined at 80° to the H.P., passing through a point 22mm from the circumference of the base.
(iii) A perpendicular plane inclined at 50° to the H.P. and passing through a point on the axis 45mm above the H.P.

CHAPTER 19

INTERPENETRATION AND SURFACE DEVELOPMENT

Interpenetration and Surface Development

In the modern industrial world, in which the pre-fabrication of welded sheet-metal parts is so important, it is necessary to be able to find the true line of intersection between two parts of an object and the true shape (or development) of a piece of sheet metal before production can be commenced.

Surface Development

Surface Development (i.e. finding the true shape of a flat surface before it is folded to form a particular object such as a metal water tank, or finding the true shape of the surface of a solid in order to determine true angles or bevels) is a practical application of the work of Chapter 12, in which the true length and angles of inclination of a line were found.

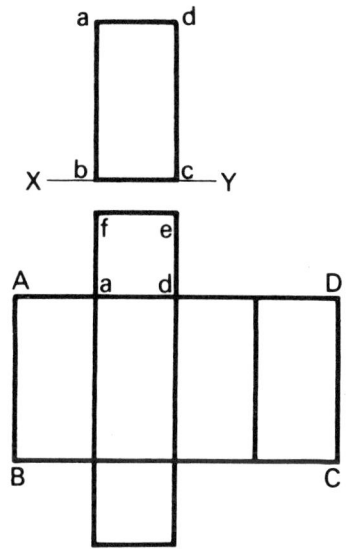

FIG. 19.1

FIG. 19.1 *To find the surface development of a right square hollow prism.*

Draw the front elevation, $a'b'c'd'$, and plan *adef*. A right square prism has six faces—two ends, of which the plan is one, and four sides of which the front elevation is one. Therefore the development can be drawn on either view. Produce *af* and *de* to make a side and end below the plan, then add three sides to make AD equal to four times $a'd'$ and AB equal to $a'b'$. At this stage it is a good plan to cut out the developed surface and fold it to form the required prism.

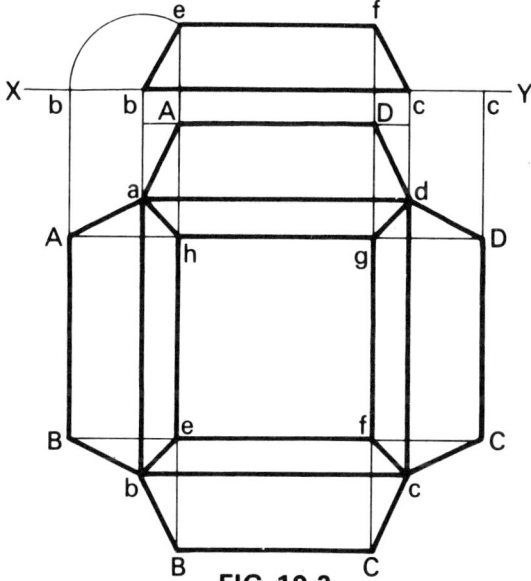

FIG. 19.2

FIG. 19.2 *To draw the surface development of a square hopper-shaped box without a lid (a box with sides that slope inwards).*

Draw the front elevation, $b'c'f'e'$, and the plan, *abcd, efgh*. Since the size of the square sheet from which the box could be formed is equal to the distance across the sloping sides and the bottom, that is $e'b' + b'c' + c'f'$, then rabat (or fold) $e'b'$ back to b'' and $c'f'$ back to c'', draw vertical projectors from b'' and c'', and draw a square about the plan of side $b''c''$. The front elevation and plan give the true lengths of the sides, therefore draw projectors from *gh, ef, fg,* and *eh* to give the sides AD, BC, CD, and AB, and join the points $ABCD$ to *abcd* to complete the development. The eight equal corner lines, e.g. *Aa,* are the only lines which are not their true length in either plan or elevation and the method used is that of Fig. 12.11.

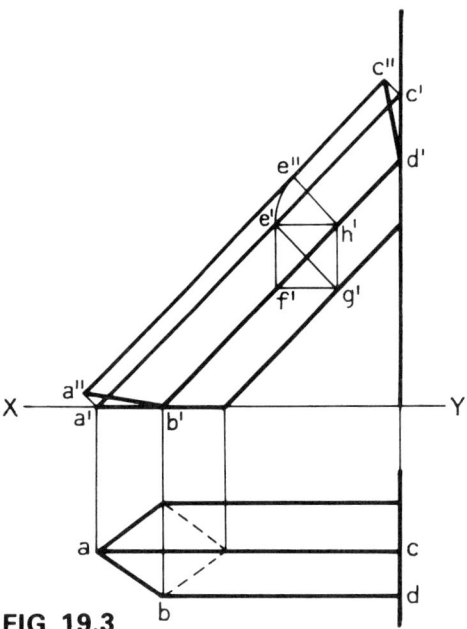

FIG. 19.3

FIG. 19.3 *To find the development of one surface of a post ABCD, given the front elevation and plan. The post is of square cross-section, fitted between a horizontal and a vertical plane, and the edges are in a vertical plane but inclined to the H.P. This could be a stay fitted between a vertical post and the ground.*

Draw the front elevation and plan. Since the post is square in cross-section, on the front elevation draw the square $e'f'g'h'$, the sides to meet on the edges of the post, one diagonal perpendicular and the other coinciding with edge $b'd'$. This square is a true cross-section of the post and the true width of surface $a'b'c'd'$ is either $e'h'$ or $e'f'$, therefore rabat $e'h'$ about h', to give $e''h'$ perpendicular to $b'd'$, and draw a parallel to $a'c'$ through e''. As in Fig. 19.2, there is no alteration in the length, so project perpendicularly from a' and c' to transfer the length $a'c'$ to $a''c''$. Join $a''b'$ and $c''d'$, then $a''b'c''d'$ is a development of one surface and gives the true shape (or bevels) for the fit between the vertical and horizontal planes.

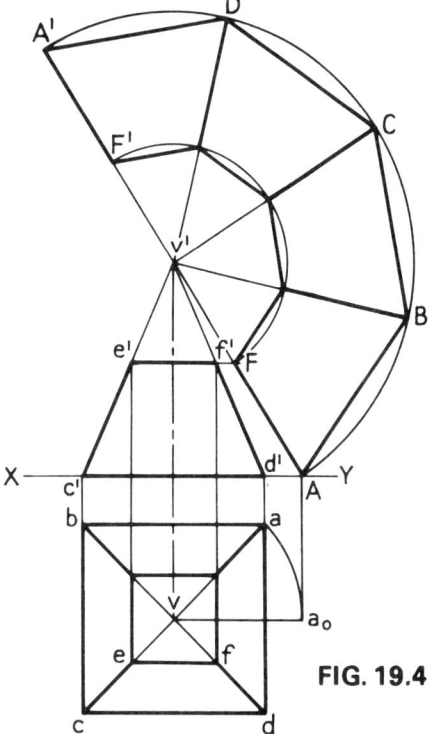

FIG. 19.4

FIG. 19.4 *To draw the surface development of a truncated right square pyramid, the basic shape of a lamp-shade.*

Draw the plan, *abcd*, and the front elevation $c'd'e'f'$. The method of development in this example is an alternative to Fig. 19.2, but still depends upon finding true lengths and angles. Find the true length of the edge va by rabating va to va_0, project a_0 on to $c'd'$ produced to give point A and join Av' the true length of the edge of the complete pyramid. To find the true length of edge on the truncated pyramid, produce $e'f'$ to F on Av', then AF is the true length of the sloping edges, $c'd'$ of the base edges, and $e'f'$ of the top edges. Describe an arc of radius $v'A$ about v', from A step off $c'd'$ four times to give AB, BC, CD, DA', and join B, C, D, and A' to v', which gives the surface development of the complete pyramid. With $v'F$ as radius centre v', describe an arc and draw four chords between the points where it cuts the lines from $ABCDA'$ to v', then $AFF'A'$ is the surface development of the truncated pyramid. This figure only gives the development of the sides, but if it were necessary to make it into a container, then the base, *abcd*, could easily be attached to one of the base edges.

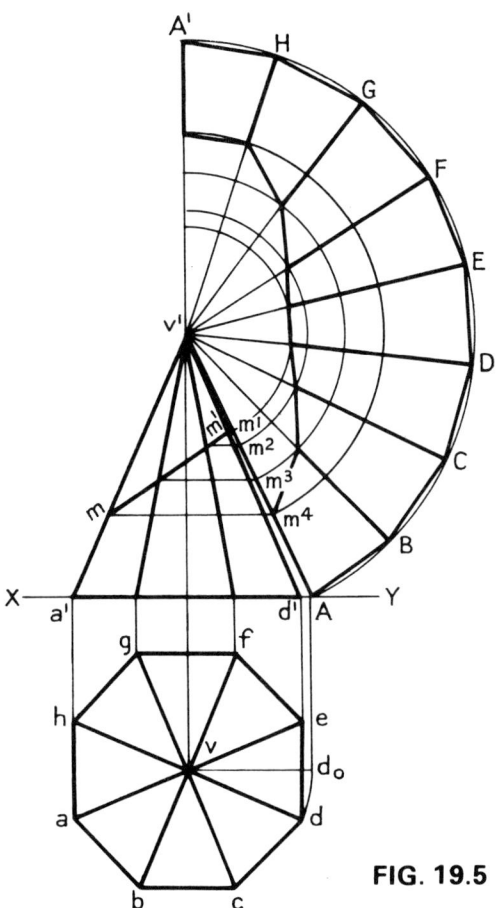

FIG. 19.5

FIG. 19.5 *To find the surface development of a truncated right octagonal pyramid. The pyramid is cut by a perpendicular plane, mm^1, inclined to the axis at a given angle.*

Draw the plan, *abcdefghv*, and the front elevation, *a'd'v'*. As in Fig. 19.4, find the true length of *vd* by rabating *vd* to *vd$_0$*. Project *d$_0$* on to *a'd'* produced to give point *A*, then *Av'* is the true length of the edge of the complete pyramid. With *v'* as centre and radius *Av'*, describe an arc of a convenient length, and from *A* step off the chord distances *AB, BC, CD, DE, EF, FG, GH*, and *HA'*, each equal to one side of the base of the pyramid, e.g. *ab*. Join *B, C, D, E, F, G, H*, and *A'* to *v'* to give the surface development of the complete pyramid. Since *Av'* is the true length of an edge, draw horizontal projectors from the points of intersection of *mm^1* with the edges to *Av'*, to give points *m^1, m^2, m^3*, and *m^4* on the true length of the edge *Av'*. With *v'* as centre, radii *v'm^1, v'm^2, v'm^3*, and *v'm^4*, describe arcs on the complete surface development and join the points of intersection on the edges to complete the surface development.

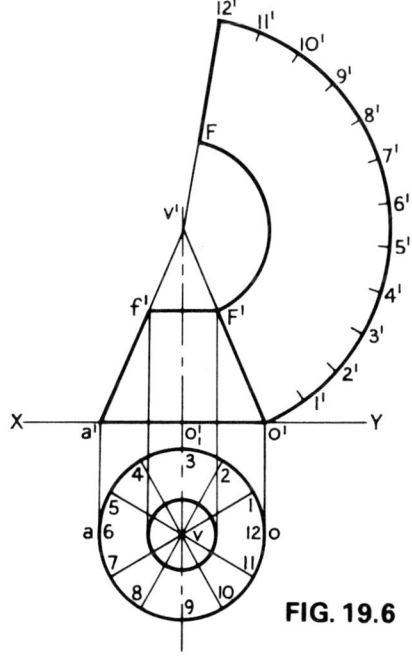

FIG. 19.6

FIG. 19.6 *To draw the surface development of a truncated right cone, a shape used for a lampshade or part of a hopper.*

Draw the plan, *av*, and the front elevation *a'o'v'*. The true length of the sloping side is given in the front elevation and the circumference of the base can be found by dividing the plan into a convenient number of equal divisions and measuring the chord of each small arc. With *v'* as centre and radius *o'v'*, describe an arc, divide the plan into twelve equal parts and number each part. On the arc *o'v'* step off twelve chords from *o'* each equal to the corresponding chord on the plan—because the chords on the plan are all equal, only one need be measured—and join *12'v'* giving *v'12'o'*, the surface development of the complete cone. Since the distance *v'F'* is true, then with centre *v'* and radius *v'F'* draw an arc to cut off the surface development of a truncated cone, *o'12'FF'*. It should be noted that the greater the number of divisions there are in the circle, the nearer the arc will be to the true circumference.

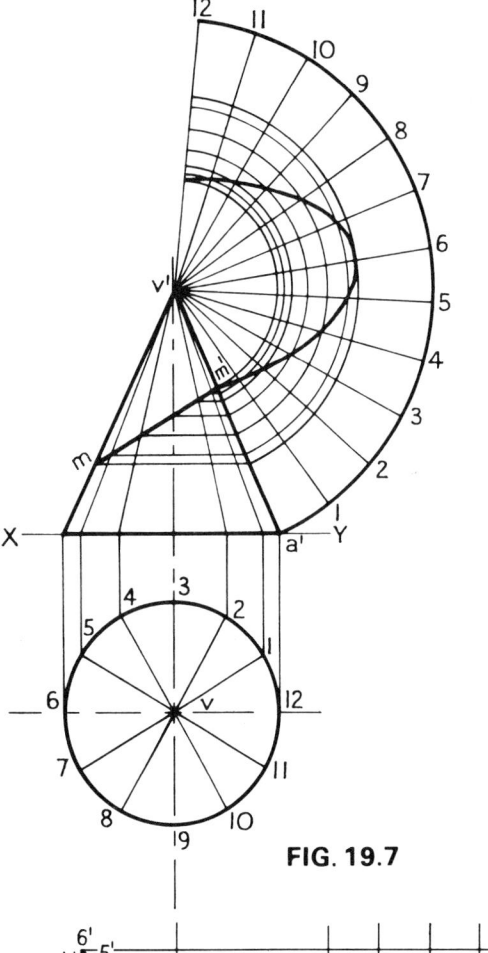

FIG. 19.7 *To draw the surface development of a truncated right cone cut by a perpendicular plane inclined to the axis at a given angle.*

Draw the plan *av* and the front elevation *a'v'*. As in Fig. 19.6, draw the surface development of the complete cone by describing an arc of radius *v'a'* from *v'*, dividing the plan into twelve equal parts and stepping off the chord lengths on the arc and joining *v'*12. Since a cone can be regarded as having an infinite number of edges on its surface, all radiating from the vertex *v'*, then the twelve lines on the plan can be regarded as imaginary edges and projected on to the elevation. As in Fig. 19.5, draw horizontal projectors from the points of intersection of *mm'* with these imaginary edges to give points on the edge *a'v'*. With *v'* as centre and radii *v'* to each of the points on *a'v'*, draw arcs on the complete surface development of the cone and then a smooth curve through the points of intersection to give the surface development of the truncated cone.

FIG. 19.7

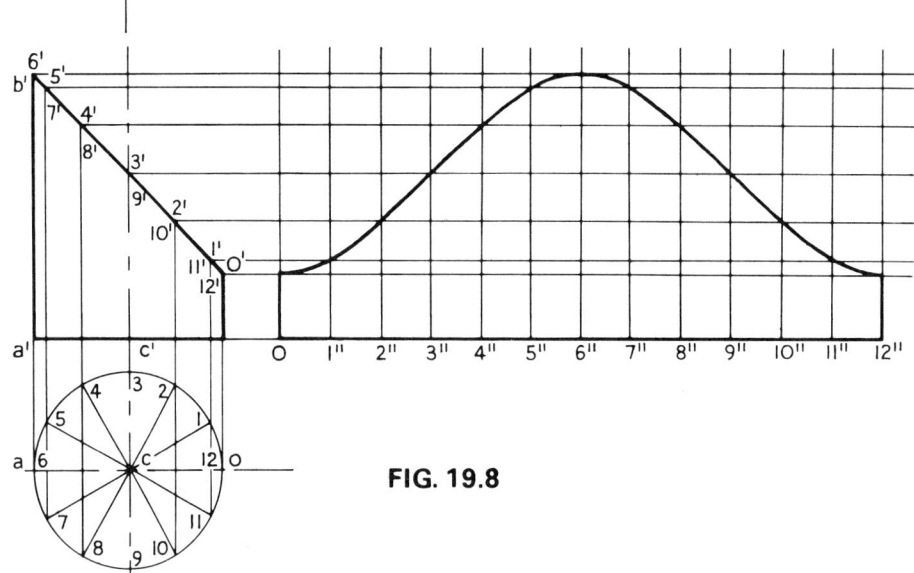

FIG. 19.8

FIG. 19.8 *To draw the surface development of a right cylinder cut at an angle of 45° (e.g. two pipes of equal diameter, brazed or welded to form a right-angled bend).*

Draw the plan and front elevation, *ac* and *a'c'b'*. The surface development of the base of the cylinder is equal to the circumference. Therefore produce *a'c'*, make *o*12″ equal to the circumference (π × the diameter of the cylinder), by means of a scale divide it into twelve equal parts, number and erect verticals at each point of division. Divide the plan into the same number of equal parts and number. Draw projectors to the front elevation to give 1′2′3′4′5′ and horizontal projectors from these points and the end of the 45° cut. By drawing a smooth curve through the points of intersection of verticals and horizontals of the same number, the surface development is completed. In this example the projectors through the plan and elevation can be regarded as vertical cutting planes with their h.t.s on the plan.

Interpenetration

Interpenetration is the word describing the intersection of two or more solids, and problems on it involve finding the true shape on the line of such intersections. The methods are those used in the preceding chapters, i.e. the projection of points of intersection and the use of auxiliary (or imaginary) cutting planes. The number of cutting planes used need not be large, but it is most important that they should be placed (i) in positions where the sections will give either straight lines or circles, (ii) on the extremities of intersections and other similar limiting positions.

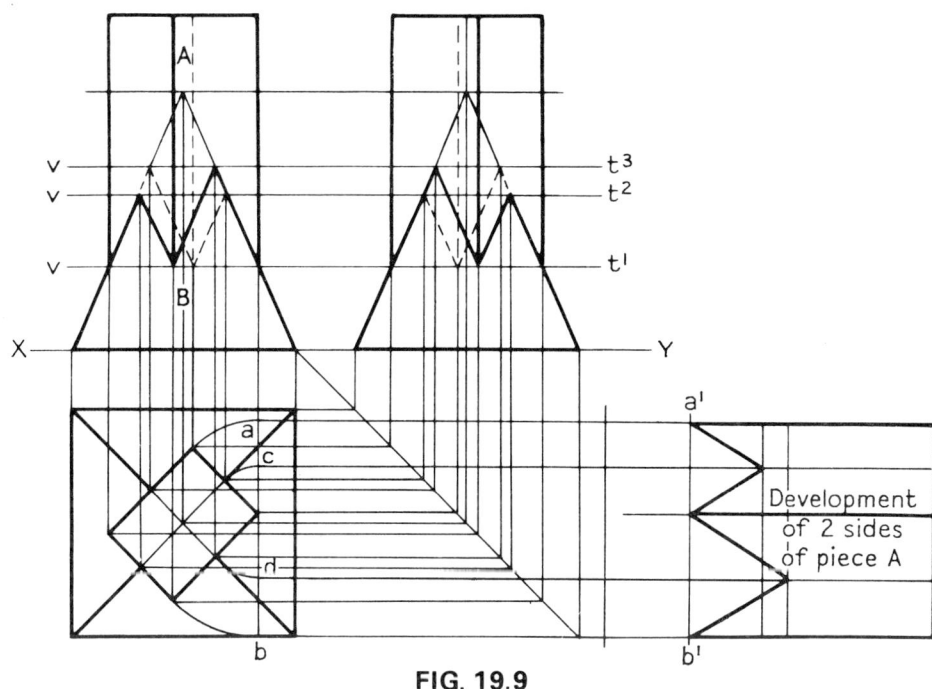

FIG. 19.9

FIG. 19.9 *To find the lines of intersection of a right rectangular prism A and a right square pyramid B. The prism and pyramid have a common vertical axis and, in plan, the sides of the prism are perpendicular to the edges of the pyramid.*

Draw the plan and project the front and end elevations of the complete pyramid and prism. Draw vertical projectors from the points on the plan where the prism cuts the edges of the pyramid and, where these projectors cut the edges of the pyramid on the front elevation, draw the vertical traces of the horizontal cutting planes, vt^1, vt^2, and vt^3. On these traces will be points on the lines of intersection between the prism and pyramid; join these points in full and hidden detail lines to complete the interpenetration on the front elevation. Draw horizontal projectors from the points on the plan, where the prism cuts the edges of the pyramid and project them on to the end elevation. Through the points of intersection complete the interpenetration on the end elevation. After interpenetration it is sometimes necessary to develop surfaces to show either the true shape of a hole or of an intersecting surface. In this example two surfaces of the prism A have been developed by the rabatment of four points on the plan of the prism to a,b,c, and d. To complete this surface development, draw horizontal projectors from a, b, c, and d and a vertical line $a'b'$ in a convenient position. Step off the vertical distances of the v.t.s above XY and transfer them to the projectors beyond $a'b'$. Join the points and complete the development.

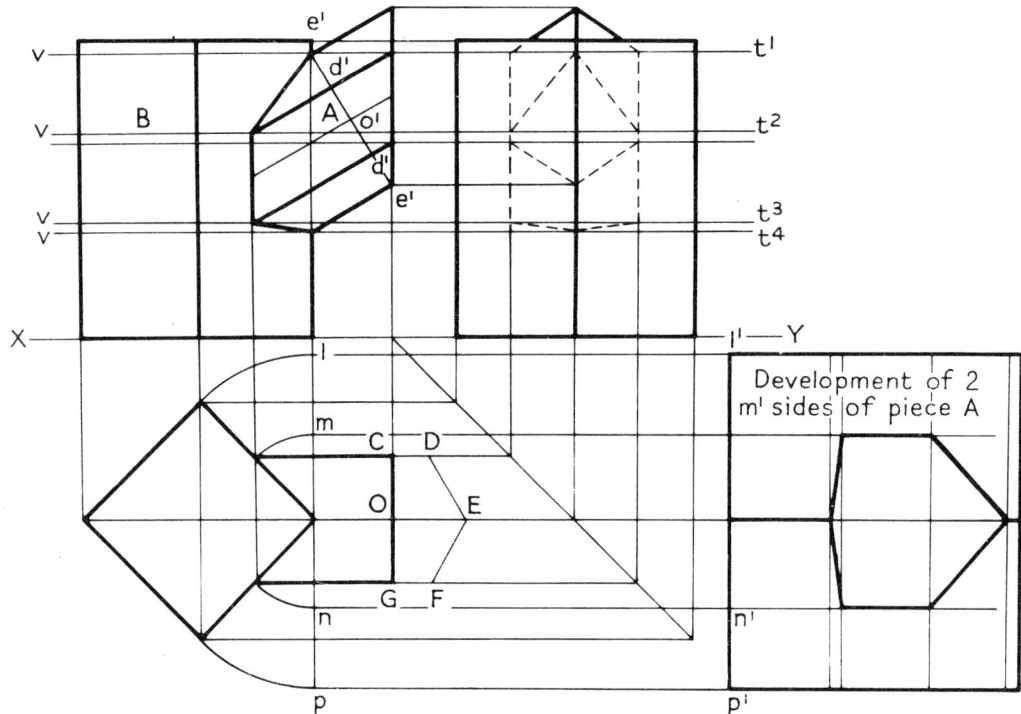

FIG. 19.10

FIG. 19.10 *To find the lines of intersection between a right prism B and part of a right regular hexagonal prism A. The axis of B is vertical, A is inclined to the vertical at a given angle, two faces are each in a vertical plane, and the top and bottom edges of A meet the same edge of B. The plan of the prisms is given.*

   Draw the plan of the square prism B, on its centre line a half cross-section of the hexagonal prism A (additional view *CDEFGO*), and complete the plan of both prisms. Draw projectors from the plan to both vertical planes and draw the front and end elevations of prism B. Draw the centre line of prism A, at the given angle, on the front elevation, and through a point o' draw a perpendicular to the centre line. Since A is part of a right regular hexagonal prism, then transfer the distances CD and OE from the additional view to the new perpendicular making $o'e'$ equal to OE and $o'd'$ equal to CD, to give four points about o'. Through the four points, e', d', d', and e', draw parallels to the centre line, on which is o', then these four lines are the edges in elevation of prism A. Draw a vertical projector from the points where the prism A meets prism B on the plan, and through points of intersection of this projector and the edge of prism B with the edge of prism A, draw the vertical traces of imaginary horizontal planes, vt^1, vt^2, vt^3, and vt^4, on the front elevation and produce them to the end elevation (as in Fig. 19.9). Draw projectors from the plan of prism A to the end elevation, and join the points of intersection in both elevations to complete the interpenetration.

   In an example of this kind it might be necessary to find the true shape of the hole in the surfaces of prism B. As in Fig. 19.9, rabat the points on two edges of the plan to l, m, n, and p, and draw horizontal projectors beyond the vertical $l'p'$. Step off the vertical distances above XY of vt^1, vt^2, vt^3, and vt^4 beyond $l'p'$ and join the points of intersection to give the true shape of the hole. From l' mark off the length of prism B and complete the rectangle containing two true surfaces of B.

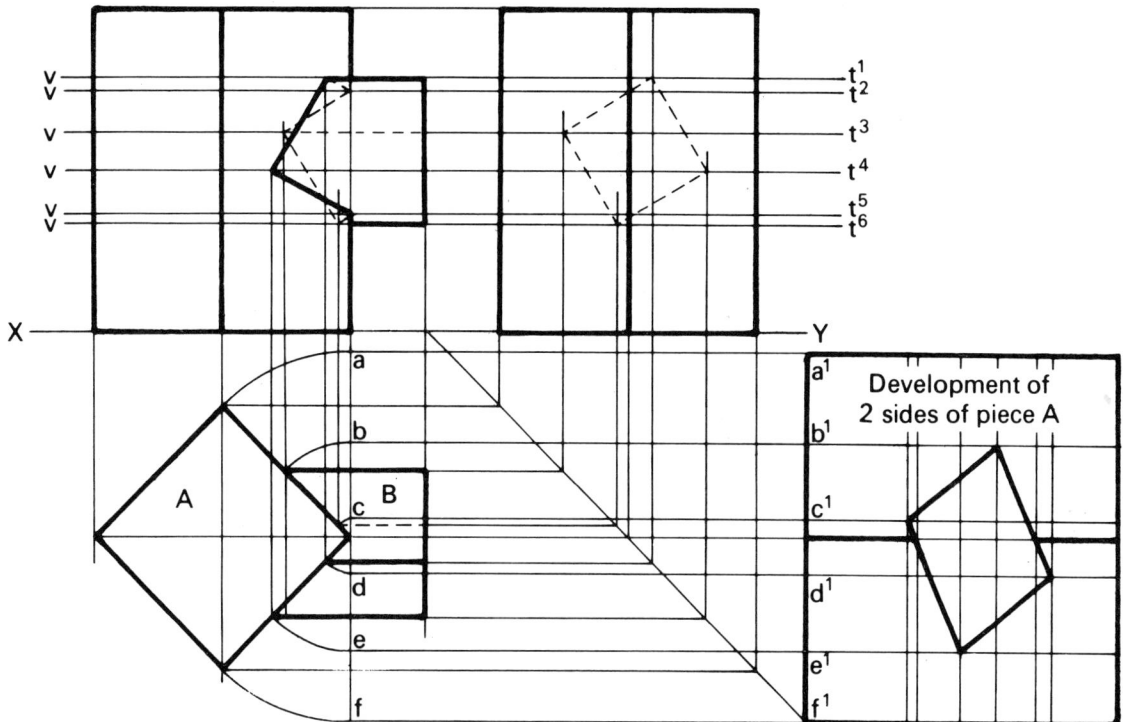

Development of
2 sides of piece A

FIG 19.11

FIG. 19.11 *To find the line of intersection between two right square prisms, A and B. A is greater than B in cross-section, the axis of A is vertical, the axis of B is horizontal, and given is an end elevation.*

Draw the plan of prism A and from it project the front and end elevations of A. Complete the end elevation of the two prisms by showing B in hidden detail. Draw projectors from B, on the end elevation, to the plan and complete the plan. Draw vertical projectors, from the points of intersection of the edges of B with those of A, from the plan to the front elevation. Draw the vertical traces vt^1, vt^2, vt^3, vt^4, vt^5, and vt^6 of imaginary horizontal planes cutting the end elevation at the points of intersection of B and A and the corners of B. On the front elevation, join the points of intersection of the traces with the projectors from the plan, to complete the interpenetration.

As in Figs. 19.9 and 19.10, a development of two surfaces can be done. Rabat points of intersection on the plan to a, b, c, d, e, and f, draw projectors from these points, a perpendicular $a'f'$, and transfer the distances above XY of the v.t.s to the projectors beyond $a'f'$. It should be noticed that in joining the intersections there are six points, four the corners of B and two the intersections of an edge of A with the cross-section of B.

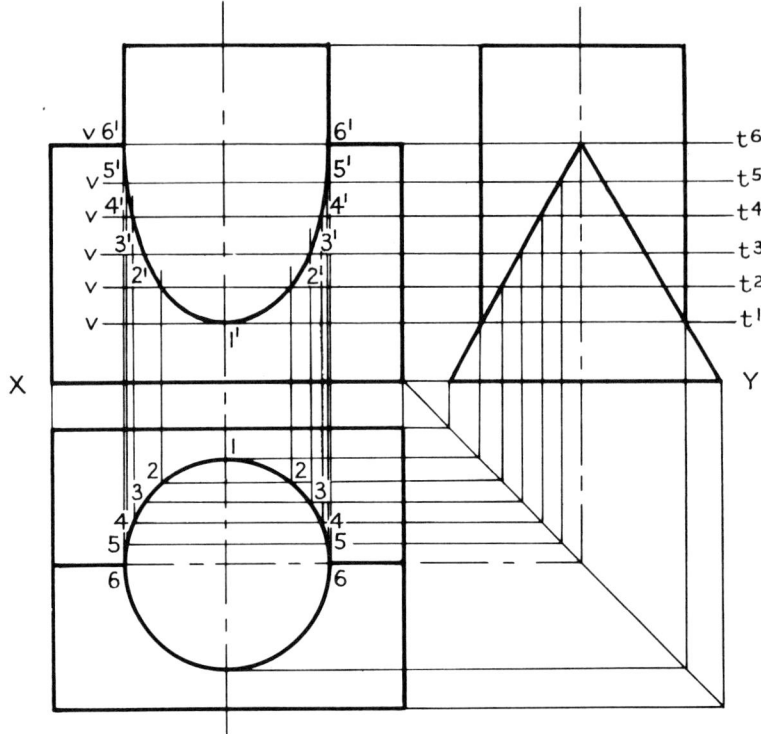

FIG. 19.12

FIG. 19.12 *To find the lines of intersection of a right cylinder and a right equilateral triangular prism of top edge greater than the diameter of the cylinder. The axis of the cylinder passes through the mid-point of the top edge of the prism and is vertical; one side of the prism is on the H.P. and parallel to the V.P.*

Draw the plan and front and end elevations of the complete prism and cylinder. The end elevation is required because it gives the limits of the line of intersection on the front elevation, therefore draw vertical traces of horizontal planes between these limits, vt^1, vt^2, vt^3, vt^4, vt^5, and vt^6. Where these traces cut the sides of the prism on the end elevation, draw projectors to the plan to cut the cylinder at points 2, 3, 4, and 5 and project to the front elevation to give points $2'$, $3'$, $4'$, and $5'$ on the v.t.s. Draw a smooth curve through points $6'$, $5'$, $4'$, $3'$, $2'$, $1'$, $2'$, $3'$, $4'$, $5'$, and $6'$ to complete the interpenetration. The true shape of the hole could be found as in Figs. 19.9, 19.10, and 19.11, but in this example it would be necessary to rabat the points on the end elevation and transfer horizontal distances from the front elevation.

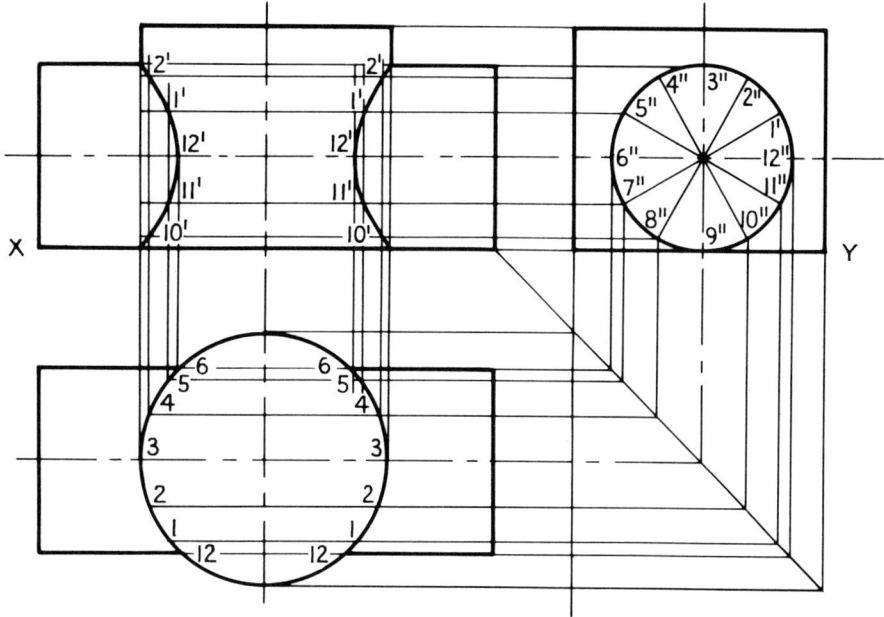

FIG. 19.13

FIG. 19.13 *To find the lines of intersection of two right cylinders of different diameter; the larger cylinder has one end on the H.P., its axis is vertical, the axis of the smaller cylinder is parallel to both planes and passes through the axis of the larger.*

Draw the plan and front and end elevations of the complete cylinders. Divide the smaller circle, on the end elevation, into twelve equal parts, as shown in Fig. 3.7. Draw projectors to the plan from these points and vertical traces through 4″, 5″, 6″, 7″, and 8″. Draw vertical projectors from 4, 5, and 6 on the plan to cut the v.t.s on the front elevation at 2′, 1′, 12′, 11′, and 10′ and join these points to complete the interpenetration. The numbers not shown on the plan and front elevation represent the hidden detail which coincides with the outline in this particular example. The surface development can be found by taking the chord length of small arc, instead of rabatment, and then drawing projectors and transferring distances as in previous figures.

A method can be used in which a semicircle is drawn on the end of the smaller cylinder and the end elevation need not be drawn. However, this should not be used until the principles of interpenetration have been learnt.

EXERCISES

1. A right square hollow prism 75mm long and 30mm square, with closed ends, has to be made from tinplate. Draw the development.

2. A square tray with sloping sides is 105mm square on the top edges, 90mm square in the bottom, has a perpendicular depth of 36mm, and has to be made from sheet brass. Draw the development.

3. A right square hollow pyramid has a 90mm axis and 45mm base. Draw the development of its sides.

4. A hollow right hexagonal pyramid is 82mm perpendicular height, 22mm base edge, and cut by:
(i) A horizontal plane 45mm above the base.
(ii) A perpendicular plane inclined at $30°$ to the base and passing through a point on the axis 22mm above the base. In each case, draw the development of the sides.

5. A right hollow cone is of 90mm axis and 75mm diameter. Draw the development of the side when it is cut by:
(i) A horizontal plane, parallel to and 60mm above the base.

(ii) A perpendicular plane inclined at $40°$ to the base and passing through a point on the axis 52mm above the base.

6. A right hollow cylinder of 38mm diameter and 82mm long is cut by a perpendicular plane inclined at $60°$ to the end and passing through a point on its axis 30mm below the top end. Draw the development of the cylinder below the plane.

7. Fig. 19.3 represents a projection (drawn to scale) of a square-section stay supporting a vertical post. The stay is inclined at $40°$ to the H.P., the length on the top edge is 120mm and the square-section is of 22mm edge. Find the true shape, and the end bevels, of the surface marked $a'b'c'd'$ in elevation.

8. A square pyramid of 105mm axis and 90mm base edge is penetrated by a right rectangular prism of end 45mm by 34mm. In plan, the edges of the prism are parallel to the corner diagonals of the pyramid with the vertex the common centre of the prism and pyramid. Draw the lines of intersection of the prism and pyramid when the pyramid is on the H.P. and both axes are vertical. Show the true shape of two surfaces of the pyramid.

9. Draw the intersection of a cylinder of 38mm diameter which penetrates a right equilateral triangular prism 90mm long and 60mm side. The plan is shown in Fig. 19.12. Develop the true shape of the hole in the surfaces of the prism.

10. A cylinder is 45mm diameter and 90mm long, is horizontal and 15mm above the H.P. It passes through a 60mm diameter cylinder which has one end on the H.P. Draw the lines of intersection where the axes intersect.

CHAPTER 20

HELICAL CURVES

Screw threads and coil springs are used in many industries and drawings of them are known as helical curves. If a piece of wire were wound evenly round a metal rod so that one end finished higher up the rod than the other, and if this wire were then welded or brazed into position, a screw thread or helical curve would have been formed.

EXERCISE

Coil a short length of thin copper wire round a pencil (as for wiring up an electrical experiment), remove the pencil and stretch the wire slightly. Four important facts emerge:

 (i) The wire was straight before being coiled.
 (ii) In coiling, one end was raised above the other to form an inclined plane.
(iii) The development of the coil or helix must be a straight line, with one end raised above the other.
(iv) The helix is the locus of a point which moves at a constant rate round and along the length of a right circular cylinder.

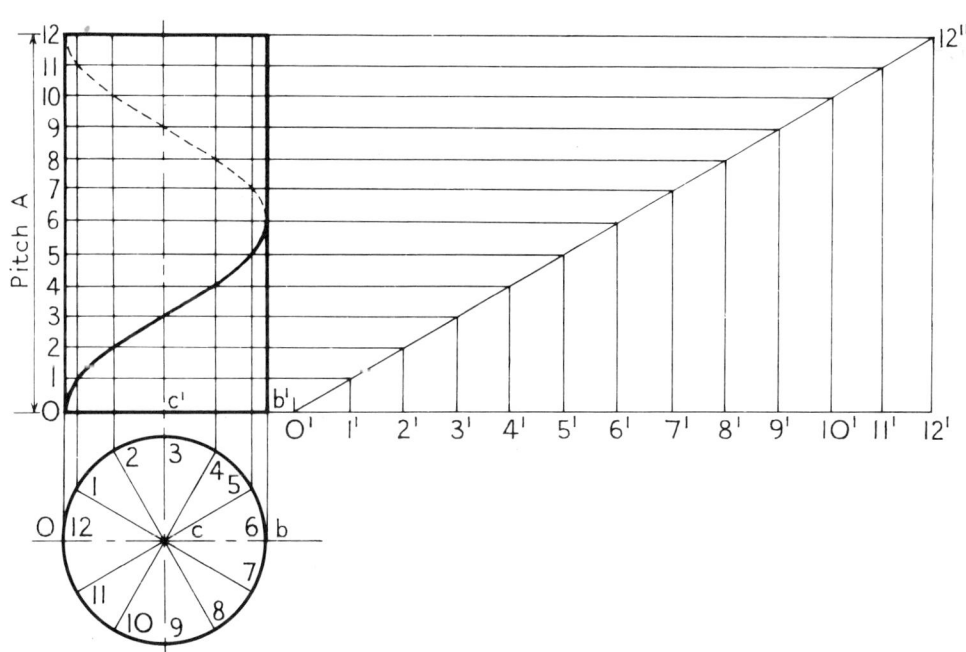

FIG. 20.1

FIG. 20.1 *To draw one complete turn of a helix of pitch distance A and radius CB.*

The pitch is the distance of rise in a complete turn, therefore draw the plan and front elevation equal to the given radius and pitch. Draw the development of the helix by making $o'12'$ equal to the circumference of the cylinder, the vertical rise $12'12''$ equal to the pitch and joining $o'12''$ to give the true length of the helix. Divide $o'12'$ into twelve equal parts, as in Fig. 19.8, the plan into the same number of equal parts, and draw projectors from the points on the circumference of the plan, 1, 2, 3, 4, and 5, to the front elevation. Erect verticals at each division on $o'12'$ and draw horizontal projectors from the points where these cut the true length $o'12''$, to the front elevation. Since $o'12''$ is the length of the wire or string which will completely circle the cylinder and will rise the predetermined height, known as the pitch, the intersection of the horizontal projectors from $o'12''$ with the vertical projectors (of the same number) are points on the helical curve. Draw a smooth curve through these points and show hidden detail.

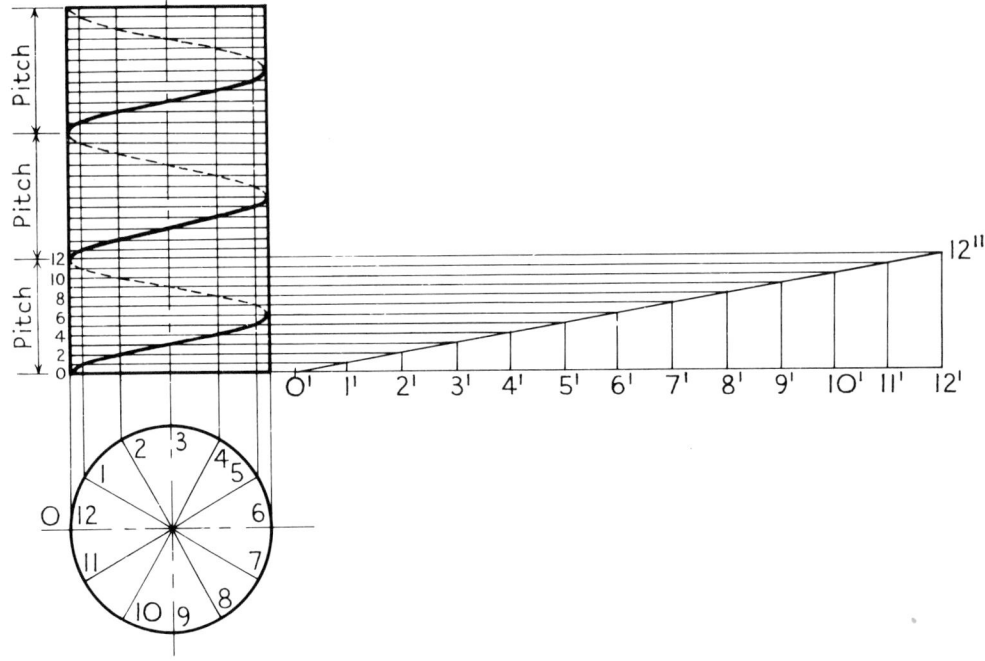

FIG. 20.2

Fig. 20.2 *To draw a helical curve of three complete turns as for a screw thread.*

Draw the plan and elevation of the cylinder about which the helix will be formed. As in Fig. 20.1, divide the plan into twelve equal parts and draw projectors from these points to the front elevation. Draw the development $o'12''$, divide $o'12'$ into the same number of equal parts as the plan and erect verticals at each point. Draw horizontal projectors from $o'12''$ to the front elevation so marking off the pitch and dividing it into twelve equal spaces. Mark off these spaces and the two pitch distances above $12'12''$ and, as in Fig. 20.1, draw a smooth curve through the points of intersection and show hidden detail.

An alternative method of finding points on the second and third turns is that of marking the pitch distance from points on the first turn and along the projectors from the plan. This method is used in Figs. 20.3, 20.4, and 20.5.

Figs. 20.1 and 20.2 give the basic principles upon which screw threads are projected, but are not complete screw threads. Fig. 20.3, 20.4, and 20.5 show more fully the application of the helix to the projection of the ISO (International Organisation for Standardisation) V thread, the square and the acme threads (both used on machine parts), respectively.

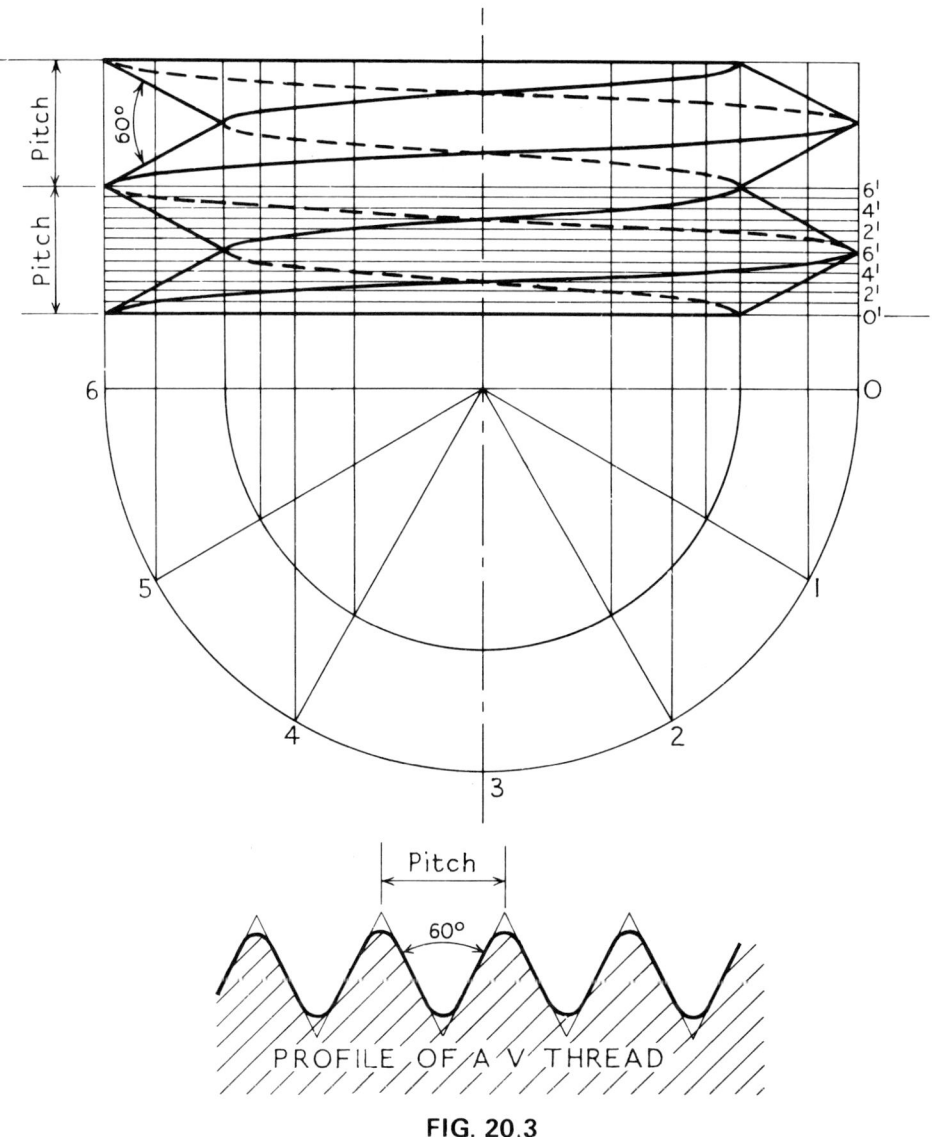

FIG. 20.3

FIG. 20.3 *To draw two complete turns of a right-hand V thread, given the diameter and pitch.*

The profile is given to show the shape of the ISO V thread. However, the required views show it as a full V (not rounded as in the profile).

Draw a half-plan and a front elevation of a cylinder of the given diameter and of axis equal to twice the pitch. Divide the axis by drawing a horizontal line a pitch distance above *XY* and draw the V shape of the thread, as in the profile but not rounded. This V determines the smaller diameter (or core diameter) of the thread, therefore project the point of the V to the half-plan and describe the second semicircle, showing the core diameter. Divide the half-plan into six equal parts (i.e. equal to twelve on a full plan), and divide the pitch distance above *XY* into twelve equal parts giving points $0'$–$6'$. Draw horizontals through points $0'$–$6'$ and vertical projectors from the points on the circumference of both the semicircles. Through the points of intersection of the verticals from the outer semicircle and the horizontals above *XY*, draw the outer helix of one turn from the *X* end of *XY*. On the vertical projectors, and from these points, step off the pitch distance to give points on the second turn and complete the helix of two turns. On the intersection of the verticals, from the inner semicircle, and the horizontals above *XY*, draw a complete turn of the smaller helix starting from the *Y* end on *XY*. As before step off the pitch distance on the vertical projectors to find points on the second turn and complete the helix.

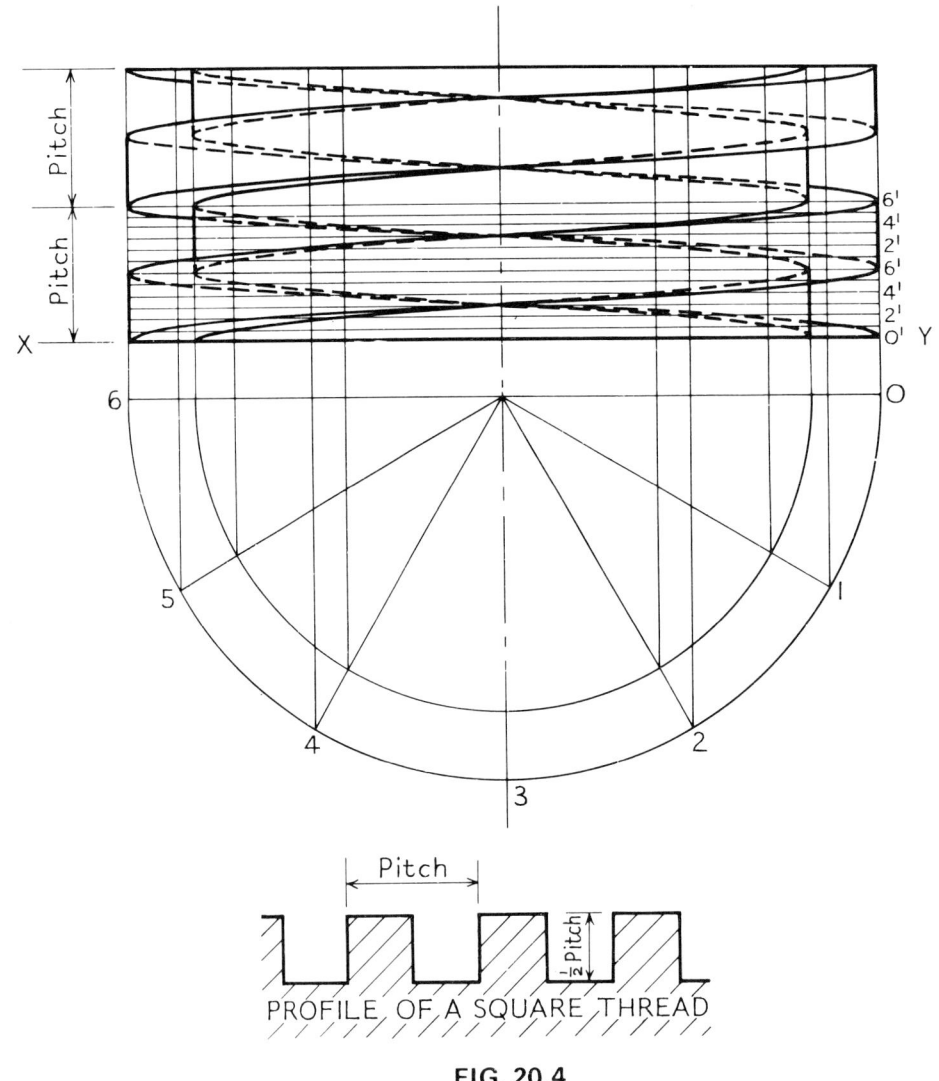

PROFILE OF A SQUARE THREAD

FIG. 20.4

FIG. 20.4 *To draw two complete turns of a right-hand square thread, given the diameter and pitch.*

The profile is given to show the shape and depth of the thread relative to the pitch.

As in Fig. 20.3, draw the half-plan and front elevation of a cylinder of the given diameter and of axis equal to twice the pitch. On the half-plan describe a second semicircle of diameter equal to the cylinder minus the pitch (i.e. the core diameter). Divide the half-plan into six equal parts (i.e. equivalent to twelve on the full plan) and draw vertical projectors from the points of intersection on the circumferences. Divide the front elevation into two pitch distances and then divide the pitch distance above XY into twelve equal parts to give points $0'-6'$. Draw horizontals through the twelve points and draw a helix of one turn, as in Fig. 20.3, through the points of intersection of the projectors from the outer semi-circle, then a second turn by stepping off pitch distances on the vertical projectors. Draw a second helix of the outer circle starting from the Y end of XY. These two helical curves are those of the full diameter of the thread and therefore can be completed with vertical lines to show the thread shape. On the intersections of the projectors from the semicircle of core diameter draw two helical curves, as for the full diameter, by starting from opposite ends of the diameter. Draw in the depth of the thread and carefully line in the hidden detail.

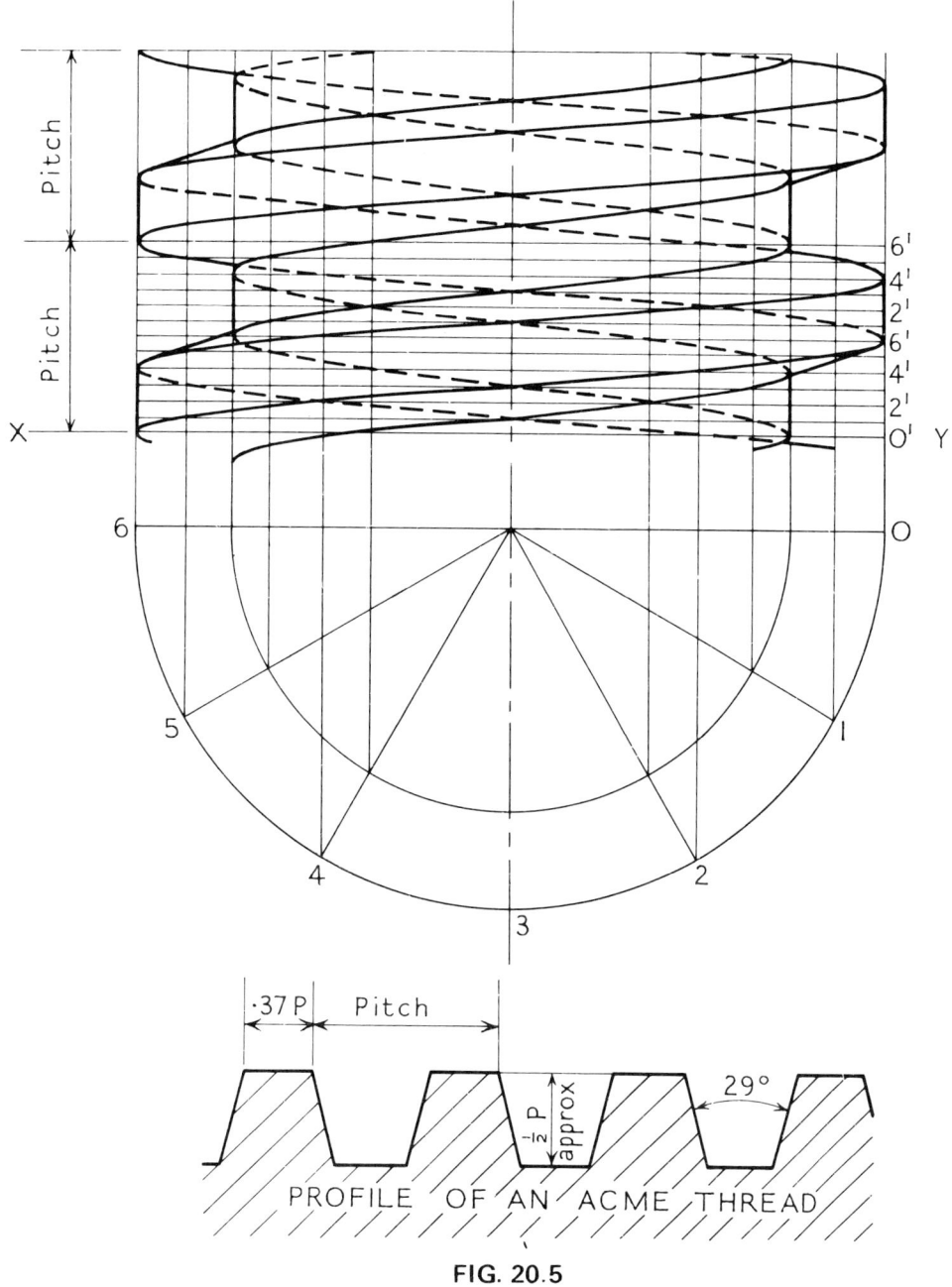

FIG. 20.5

FIG. 20.5 *To draw two complete turns of an acme thread, given the diameter and pitch.*

The profile shows the shape of this thread and the approximate relationship between the depth and pitch of the thread.

As in Figs. 20.3 and 20.4, draw a front elevation and a half-plan of a cylinder of the given diameter and of axis equal to twice the pitch. On the half-plan describe a second semicircle equal to the diameter of the thread minus the pitch (because the depth of the thread is equal to half the pitch). Divide the half-plan into six equal parts, 0–6, and draw vertical projectors from these divisions to the front elevation. Divide the front elevation into two pitch distances, the pitch distances above XY into twelve equal parts, and draw horizontal projectors through these, i.e. points $0'$–$6'$. As in Figs. 20.3 and 20.4, draw one complete turn of each of the outer helices through the points of intersection of the vertical and horizontal projectors, then, through the points obtained by stepping off a distance equal to the pitch on the vertical projectors and above points on these curves, draw the second complete turn of each of the outer helices. Draw the inner helices by the same method and line in the tops and bottoms of the thread.

EXERCISES

1. (i) A piece of string of pencil-point diameter is wrapped round a cylinder of 45mm diameter to form a helix of 60mm pitch. Draw the elevation of the cylinder showing one complete turn.
(ii) Round the same cylinder the pitch is 37mm. Draw an elevation showing three complete turns.

2. Draw two turns of a right-hand V thread of diameter 120mm and 30mm pitch.

3. Draw two turns of a right-hand square thread of 120mm diameter and 22mm pitch.

4. Draw two turns of an acme thread (right-hand) of diameter 120mm and pitch 37mm.

5. A helical spring is to be formed out of square section wire. The dimensions of the spring are 188mm long, 34mm diameter, and it has a pitch of 38mm. Draw the spring.

CHAPTER 21

ISOMETRIC AND OBLIQUE PROJECTION

It is not always easy to form immediately a picture of a completed object when first reading an orthogonal projection, but by the use of isometric or oblique projections—which are called pictorial views—this difficulty can be overcome. Neither of these projections can be regarded as complete working drawings, because some if not all views of the surfaces are distorted, but the lengths of horizontal and vertical lines are the same as in an orthogonal projection and measurements can therefore be taken from them.

Isometric Projection

Isometric projections are drawn with all horizontal lines at 30° to the horizontal, all vertical lines vertical, and Fig. 21.1 shows the three isometric axes which form the basis of any isometric drawing. In Fig. 21.2 a rectangular block is drawn showing that all parallel lines on the object are parallel in isometric and hidden detail lines are used as in orthographic, but these latter are not always essential, because the drawing is generally a form of auxiliary projection.

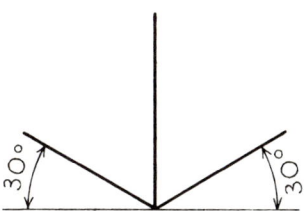

FIG. 21.1

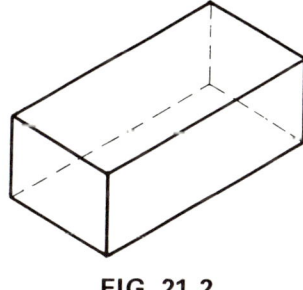

FIG. 21.2

Fig. 21.3 *To draw a wedge in isometric projection.*

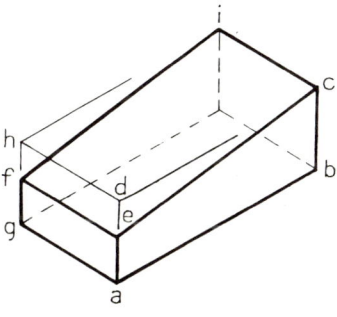

FIG. 21.3

The method introduced in this example is known as 'boxing in'. This is necessary for most pictorial views and is a very important principle. A sculptor generally starts with a rectangular block of stone from which he removes pieces to produce the shape he requires. Boxing in is very similar. Draw the vertical at *a*, mark *d*, and draw lines at 30° to the horizontal to *b*, *c*, *g* and *h;* complete the rectangular block, *abcdghi*, and mark off the thin end from point *e*. Join *e* to *c*, *f* to *i*, and line in. The top part of the block has been removed leaving the required wedge.

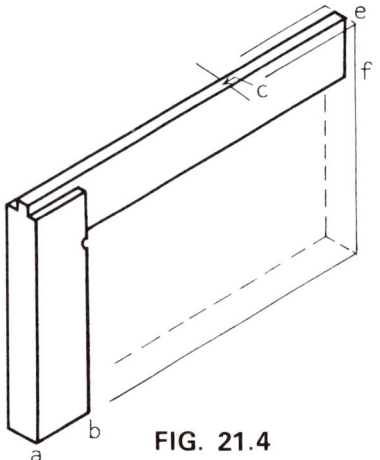

FIG. 21.4

FIG. 21.4 *To make an isometric projection of a try square.*

Using the method of Fig. 21.3, draw a block in isometric projection equal to the overall dimensions of the try square. In the middle of the top edge draw the thickness of the blade, *c*, and from *c* its width *ef*. Draw the width of the stock *ab* and its length and distance below the top edge of the blade. This cutting by pencil lines removes a rectangular block between the stock and the blade, pieces each side of the blade, and, after lining in, gives a clear outline of the try square.

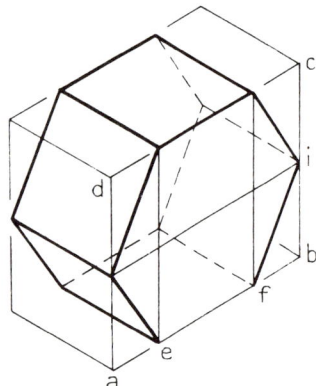

FIG. 21.5a

FIG. 21.5*a* and 21.5*b* *To make an isometric projection of a regular hexagonal prism, e.g. a blank to be used in the manufacture of a nut.*

Since lines, other than those which are vertical and horizontal, are not their true length in isometric, it will be necessary to draw an additional view of the hexagon in orthographic projection, as has been explained in the preceding chapters. Draw the additional view, Fig. 21.5*b*, *a'b'c'd'*, and a block in isometric equal to the overall dimensions of the prism. Draw vertical and horizontal lines from *e'f'*, and *i'* on the additional view to form rectangles which contain the sloping lines, and then, with the same dimensions, draw these vertical and horizontal lines in isometric on the end of the block at *e*, *f*, and *i*. Draw the hexagonal shape of one end and project the other end from it.

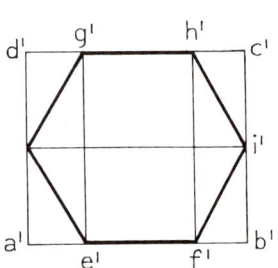

FIG. 21.5b

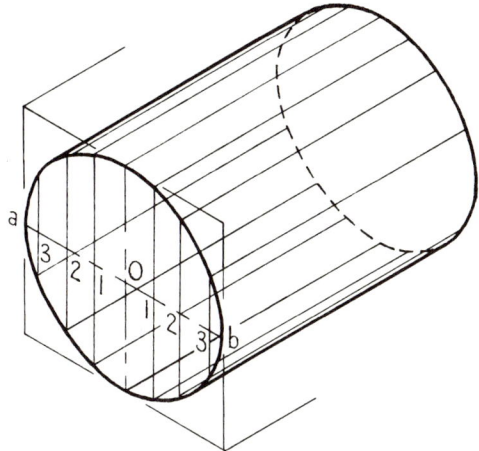

FIG. 21.6a

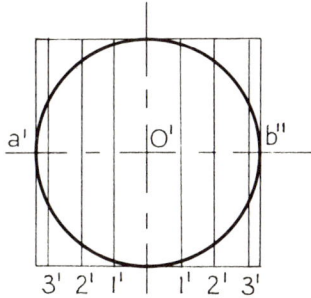

FIG. 21.6b

FIG. 21.6a and 21.6b. *To make an isometric projection of a right cylinder.*

A circle cannot be true in isometric. Therefore an additional view in orthographic projection, and within a square, must be drawn. Draw the additional view, Fig. 21.6b, and draw ordinates 1′, 2′, and 3′ both sides of the centre line. Draw centre lines, ordinates, and a square in isometric projection; transfer the lengths from Fig. 21.6b (either by stepping off from the centre line or from the edge of the square) to ordinates in the isometric view and join the points with a smooth curve through *a* and *b*. Draw horizontals, in isometric projection; then from these points make each horizontal projector equal to the length of the cylinder and draw the other end. Draw common tangents to the two end curves to complete the projection of the cylinder.

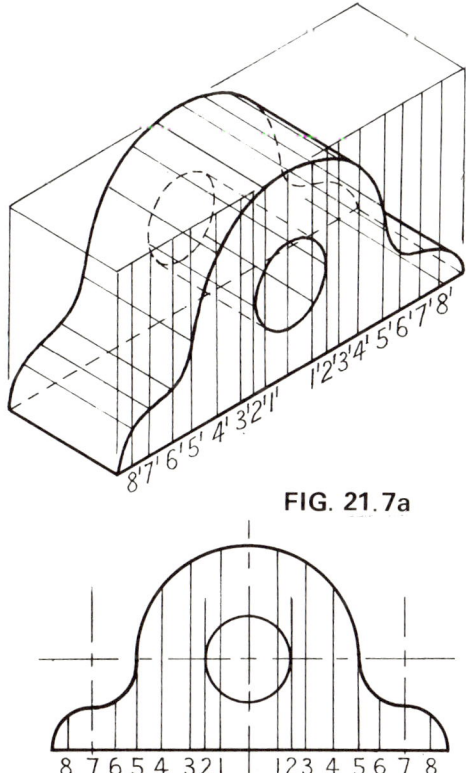

FIG. 21.7a

FIG. 21.7b

FIG. 21.7a and 21.7b *To make an isometric projection of a solid of double curvature, e.g. a casting of a simple journal bearing or a shaped clockcase.*

As in Fig. 21.6, an additional view will be necessary, therefore draw a front elevation of the solid. On this elevation draw vertical ordinates both sides of the centre line and at the extremities of curves. Draw centre lines, ordinates, and a rectangular block to contain the solid in isometric projection. Since the bottom edge of the solid is a straight and horizontal line, all the ordinate distances can be taken from this line, so step off these distances and transfer them to the corresponding ordinates on the isometric projection. Draw smooth freehand curves through the points on the isometric projection to complete one side. Draw horizontals, in isometric projection, from points on the curve, then measure off on each horizontal a distance equal to that across one end of the solid and, through these points, draw a second curve to complete the isometric projection. Join the two curves where necessary, show hidden detail, and line in the projection.

Oblique Projection

Oblique projections are similar to isometric: (*a*) they represent the complete object, (*b*) they must be drawn by using overall dimensions and then cutting away or removing pieces to form the shape, (*c*) auxiliary views and ordinates are sometimes necessary to draw angles and curves, (*d*) all vertical lines are vertical and the lengths of lines are the same as those in an orthographic projection.

The differences are that one set of horizontal lines is horizontal and the other set is usually drawn at 45° to the horizontal. Therefore there are two positions in which each object can be drawn. Figs. 21.8*a* and 21.8*b* show the axes.

Fig. 21.9 shows the rectangular block of Fig. 21.2, drawn in oblique projection. Again all parallel lines are drawn parallel and dimensions can be taken from the drawing.

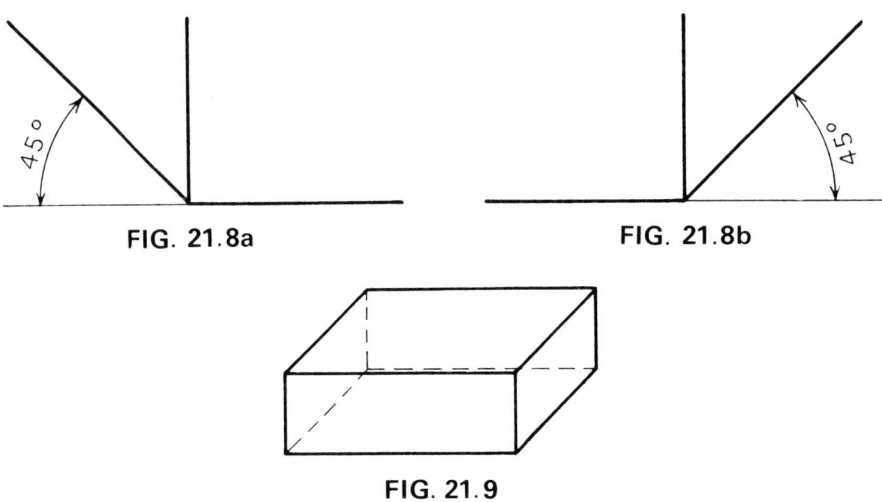

FIG. 21.8a   **FIG. 21.8b**

FIG. 21.9

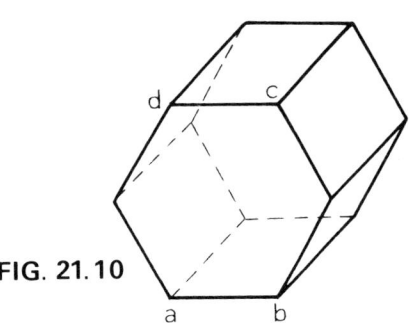

FIG. 21.10

FIG. 21.10 *To make an oblique projection of a right hexagonal prism.*

By drawing the prism so that the horizontal lines of the end are horizontal, the end is shown as a true hexagon. Draw the end of the prism, *abcd*, and from it horizontal projectors, in oblique projection at 45°, making each equal to the length of the prism. Join the ends of these projectors to complete the projection.

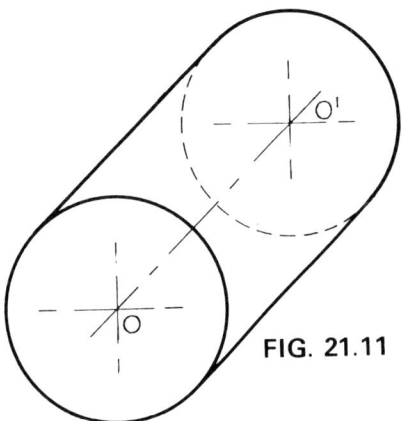

FIG. 21.11

FIG. 21.11 *To make an oblique projection of a right cylinder, the horizontal centre line of both ends horizontal.*

With the horizontal centre line horizontal the ends will be shown as true circles. Therefore draw one end, and from its centre *O* project a centre line at 45° and equal in length to that of the cylinder. From the new centre, *O'*, draw the other end of the cylinder and common tangents to the two circles.

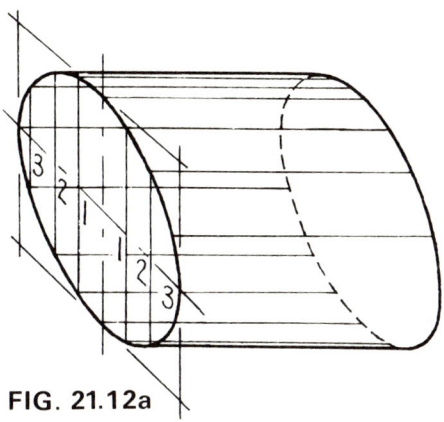

FIG. 21.12a

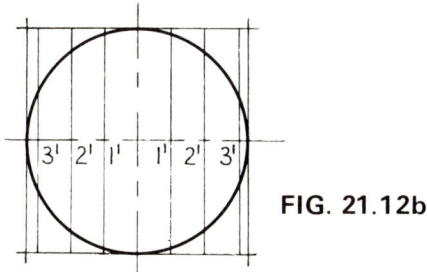

FIG. 21.12b

FIG. 21.12a. *To make an oblique projection of a right cylinder, the horizontal axis drawn horizontal.*

The ends will not be shown as true circles, and so, as in Fig. 21.6, an additional view will be necessary. Draw the additional view, Fig. 21.12b, and its ordinates. In oblique projection, draw the centre lines, the square, and the ordinates of one end and transfer the lengths of the ordinates from the true circle to the oblique view. Join the points to complete one end and from these draw projectors equal in length to the length of the cylinder. Draw the other end and common tangents to complete the projection.

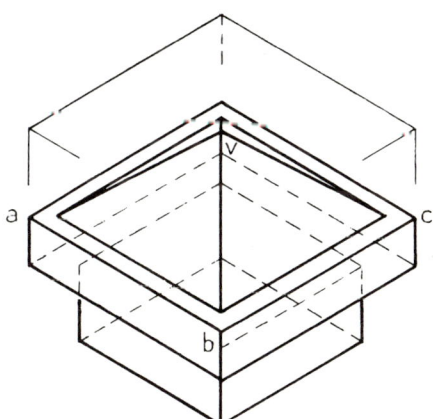

FIG. 21.13a

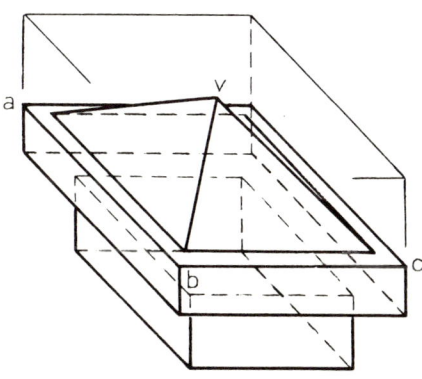

FIG. 21.13b

FIGS. 21.13a, Isometric, and 21.13b, Oblique, are projections of what could be a pyramid-shaped cap or coping of a pillar.

The method of showing each one is to draw a block equal to the overall dimensions, mark the vertex *v* in the centre of the top, draw the base of the pyramid on the plane *abc*, and join *v* to the corners. The pillar is then formed by removing rectangular sections from below the coping.

At this point, the fact that oblique axes can be inclined at other angles than 45° and shortened to a suitable scale to reduce distortion, can be introduced and applied by repeating the Exercises.

EXERCISES

Draw in isometric and oblique projection:

1. A brick 250mm by 100mm by 80mm. Scale of 1:4 full size.

2. A wedge 112mm long, 64mm wide, 45mm thick at one end tapering to 26mm thick at the other. Scale full size.

3. An octagonal right prism 90mm long and base edge 18mm. Scale full size.

4. A hexagonal blank for a nut 48mm across the flats and 30mm thick. Scale full size.

5. A pyramid on a square base of 45mm edge and 60mm axis. Scale full size.

6. A circular disk 15mm thick and 68mm diameter. Scale full size.

7. A square nut of 48mm edge by 30mm thickness with the bolt hole 30mm diameter. Scale full size.

8. The top of a model square pillar, the elevation of which is given below. Scale full size.

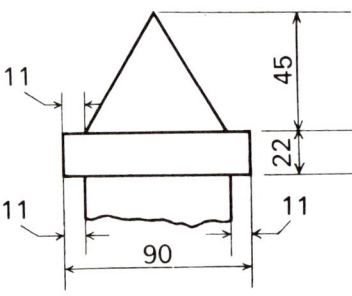

EXAMINATION QUESTIONS IN PLANE AND SOLID GEOMETRY

GENERAL CERTIFICATE EXAMINING BODY

A.B.	The Associated Examining Board.
J.M.B.	The Joint Matriculation Board.
O. & C.B.	The Oxford and Cambridge Schools Examination Board.
S.U.J.B.	The Southern Universities' Joint Board.
U.C.	The University of Cambridge Local Examinations Syndicate.
U.L.	The University of London.
U.O.	The University of Oxford Local Examinations.
W.J.C.	The Welsh Joint Education Committee.

B. PLANE GEOMETRY

Test 1

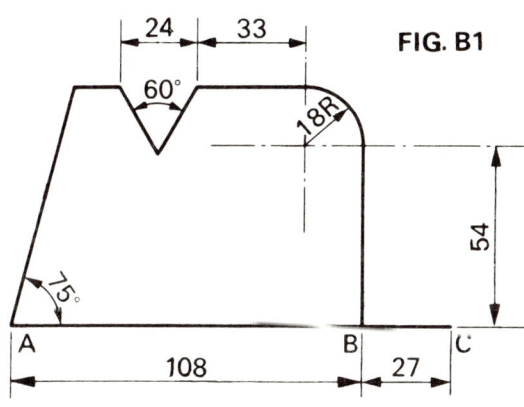

FIG. B1

1 (S.U.J.B., 1965).

(a) A line 111mm long represents a distance of 144mm. Construct a scale on this line to read in 3 millimetres.

(b) Draw Fig. B1, taking its dimensions from the scale you have constructed. Produce AB to C and on AC construct an enlarged similar figure.

2 (U.O., 1969). A triangle ABC has a base AB of length 10cm and the angle at C is 60°. If the length of BC is 0·6 times the length of AC, construct the triangle. Measure and state the lengths of the sides AC and BC. Alongside this triangle construct an isosceles triangle of equal area standing on a base of length 8cm and within this triangle draw two equal circles touching each other and the two sides of the triangle.

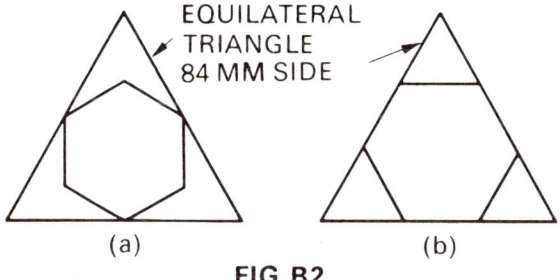

EQUILATERAL TRIANGLE 84 MM SIDE

(a)   (b)

FIG. B2

3 (W.J.C., 1965)

(i) Construct a regular hexagon with sides 72mm long,

(ii) Fig. B2 (a) and (b), show regular hexagons inscribed symmetrically in equilateral triangles. Draw the figures as shown and dimension neatly the length of side which you obtain for each hexagon.

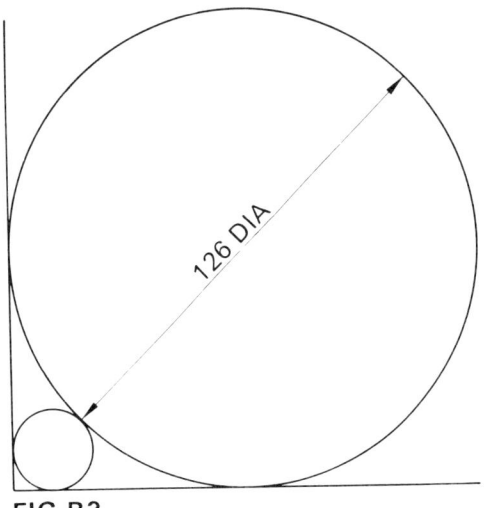

FIG. B3

4 (U.L., 1966). Fig. B3 shows two touching circles placed in the corner made by two lines which are perpendicular to one another. Draw the view shown and state the diameter of the small circle. Your construction must show clearly the method of obtaining the centre of the small circle.

5 (O. & C.B., 1970). The outline of a mechanism is shown in Fig. B4. From the initial position shown, in which the lever *BD* is vertical, the end *B* slides horizontally to position *A*.

During the movement the lever rests always on the surface of a cylinder, shown as a circle.

Whilst the end *B* moves to position *A*, a further point of the mechanism *F*, initially at *D*, moves to *C*, both motions being uniform.

Plot, full size, the locus of *F* for a complete movement of *BD*.

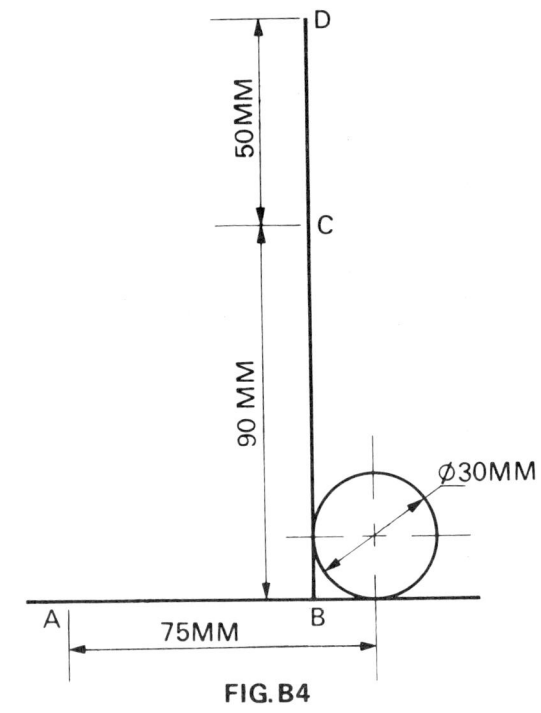

FIG. B4

6 (O. & C.B., 1965). An arch has a span of 36m and a central rise of 12m, and the centre line is an arc of a parabola. Draw the centre line of the arch to a scale of 1:250.

Test 2

1 (O. & C.B., 1969). Construct a diagonal scale on which 100mm represents 1000mm. The maximum length required off the scale is one which represents 1000mm; the smallest unit required is one which represents 1mm. Mark on the scale a length which represents 207mm.

2 (A.B., 1966). The diagonals of a rhombus measure 114mm and 22mm respectively.

(*a*) Construct, full size, the rhombus and then draw a circle to pass through both ends of the shorter diagonal and one end of the longer diagonal.

(*b*) Redraw the rhombus and by geometrical construction draw a square having an area equal to that of the rhombus. Measure and state the length of the side of the square to the nearest millimetre.

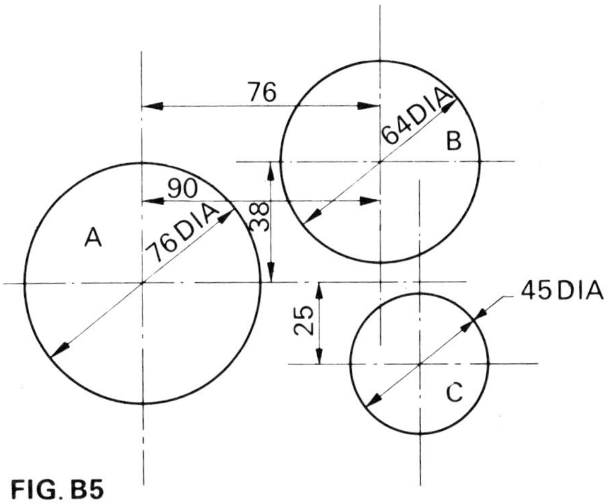

FIG. B5

3 (U.C., 1965). Three circles lie in a plane in the position shown in Fig. B5. Draw the given figure and plot the locus of a point which moves in the plane so that it is always equidistant from the circumferences of the circles A and B.

Plot also the locus of a point which moves in like manner between the circles A and C.

Finally draw a circle whose circumference touches the circles A, B, and C and measure and state its diameter.

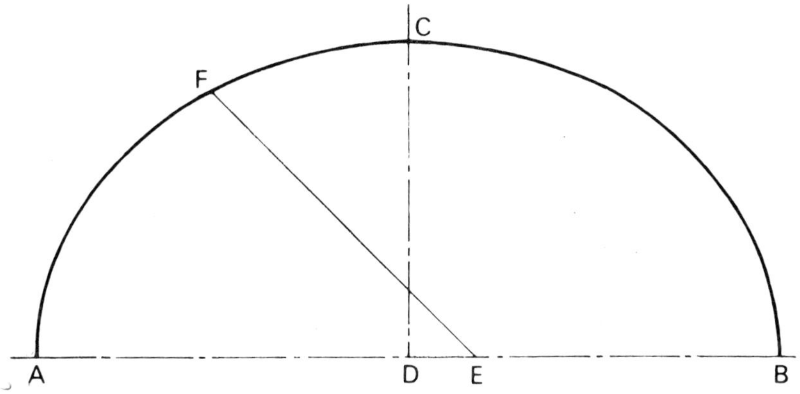

FIG.B6

4 (U.L., 1966). A semi-ellipse has a major axis AB, of 180mm and a semi-minor axis CD, of 75mm. A line EF, 90mm long, moves so that end E is always on the major-axis and end F on the circumference. Fig. B6 shows one such position. Draw the locus of the midpoint of EF.

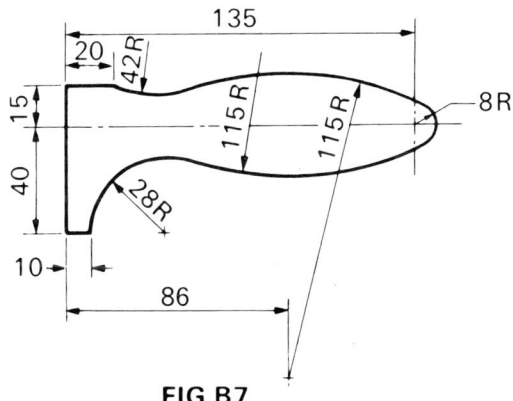

FIG. B7

5 (U.O., 1970). Set out the shape of the part of the adjustable spanner as shown in Fig. B7, showing clearly all construction and points of tangency.

Scale: full size.

If the given part of the adjustable spanner is cut out of a rectangular bar which is 20mm thick, 80mm wide and 150mm long, what is the actual volume of material used in making this shape?

Test 3

1 (A.B., 1966)

(*a*) Draw, full-size, a pentagon *ABCDE* having the following dimensions: side *AB* = 66mm, side *BC* = 57mm, side *CD* = 60mm, side *EA* = 45mm, diagonal *AC* = 96mm, diagonal *AD* = 84mm, and angle *AED* = 90°.

(*b*) Construct a similar pentagon having an area equivalent to two-thirds that of *ABCDE*.

(*c*) Find, by construction, the ratio of similar sides, expressed in the form 1:*x* where *x* is expressed to the nearest millimetre.

2 (J.M.B., 1970). Draw a 128mm diameter circle and inscribe two circles of 50mm and 38mm diameter, respectively, tangential to each other and to the 128mm diameter circle.

Construct the internal common tangent to these two smaller circles and produce it to cut the 128mm diameter circle. To one of these points of intersection construct a tangent to the 128mm circle. Measure and state the angle between the two tangents.

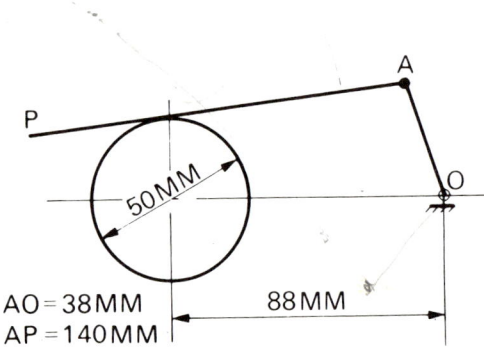

AO = 38MM
AP = 140MM   88MM

FIG. B8

3 (J.M.B., 1970). Fig. B8 shows a crank *OA* which rotates about *O*. *AP* is a rod, pin jointed at *A*, which rests on the edge of a disc of 50mm diameter as the crank rotates.

Draw the locus of end *P* of the rod during one complete revolution of crank *OA*.

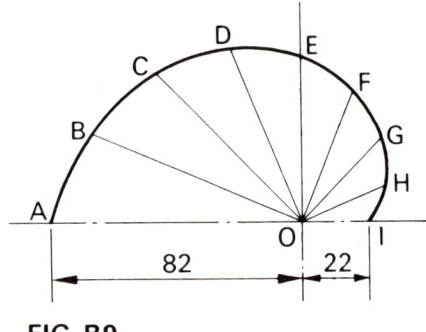

82   22

FIG. B9

4 (W.J.C., 1966). The half profile of a cam, shown in Fig. B9, is determined as follows:

(*a*) the angles between the successive lines *OA*, *OB*, *OC*, etc. are equal;

(*b*) the successive lines *OA*, *OB*, *OC*, etc., decrease in length by equal steps from *OA* = 82mm to *OI* = 22mm.

Draw

(i) the half profile as shown in the figure,

(ii) a similar half profile reduced in scale to $\frac{3}{4}$ of the original size.

5 (U.C., 1965). A vertical line and a fixed point, 25mm from it, lie in a plane. A point moves in the plane such that the ratio of its distances from the vertical line and the fixed point is 8:9. Construct the curve which depicts the path of the moving point. Make the distance of the ends of the curve approximately 85mm from the fixed point.

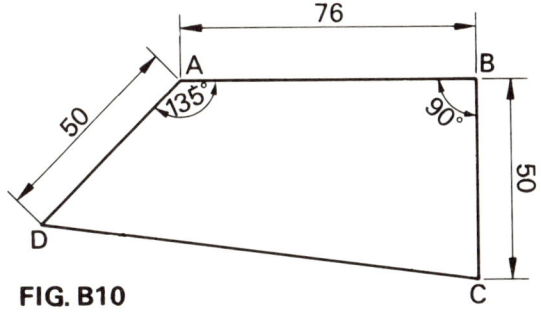

FIG. B10

Test 4

1 (U.L., 1965). At Fig. B10 is shown a quadrilateral *ABCD*. Construct a square having the same area as the quadrilateral and measure its side, stating your answer to the nearest 0·2mm.

2 (O. & C.B., 1969). Construct a triangle with a perimeter of 148mm and sides in the proportion of 21:31:28.

Construct a similar triangle which has a circumscribing circle of 88mm diameter.

Measure the longest side of this similar triangle and state its length to the nearest mm.

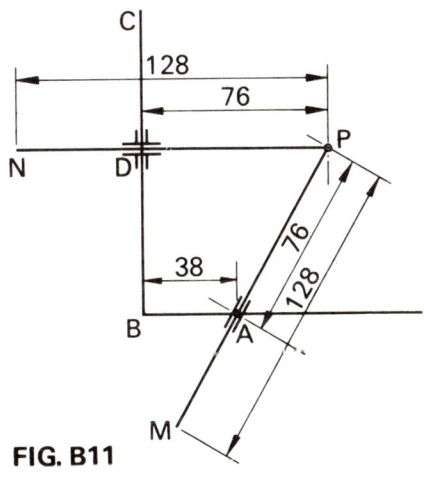

FIG. B11

3 (S.U.J.B., 1965). In Fig. B11 *MP* and *NP* are rods hinged at *P*, and *A* and *D* are guides through which *MP* and *NP* are allowed to move.

D is allowed to move along *BC*, but rod *NP* is always perpendicular to *BC*. The guide *A* is allowed to rotate about its fixed point. Draw the locus of *P* for all positions of *P* above *AB* and when *P* is always equidistant from *A* and *BC*.

This locus is part of a recognized curve. Name the curve and the parts used in its construction.

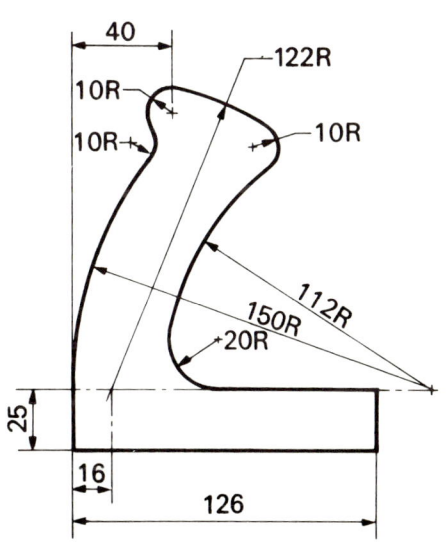

FIG. B12

4 (U.O., 1969). The outline of the handle for a hand tool is shown in Fig. B12. Draw this outline and beneath it print neatly the title 'OUTLINE OF HANDLE'.

Scale: full size.

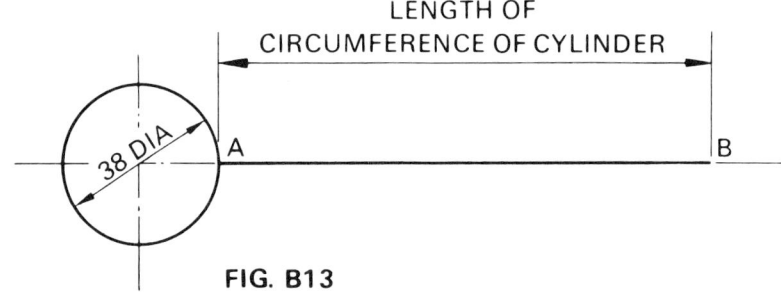

FIG. B13

5 (A.B., 1965). Fig. B13 shows a string *AB* attached to a cylinder, the length of *AB* being equal to the length of the circumference of the cylinder. Calculate length *AB* and plot the path of the end *B* as the string is kept taut and wound clockwise on the cylinder.

Test 5

1 (J.M.B., 1970). Using only a rule, a pair of compasses and a straight-edge construct sufficient of a scale of chords and then the irregular polygon *ABCDEF* to the following dimensions:

Lengths		Angles (measured clockwise)	
AB	*BC* 50mm	*ABF*	30°
BF	79mm	*ABE*	75°
BE	*BD* 90mm	*ABD*	112½°
		ABC	135°

Construct a perpendicular from *E* to *AB* and measure and state its length.

2 (U.L., 1966). The lengths of the sides of a square and an equilateral triangle are 76mm and 64mm respectively. By geometric construction draw another square equal in area to the sum of the areas of the other two figures.

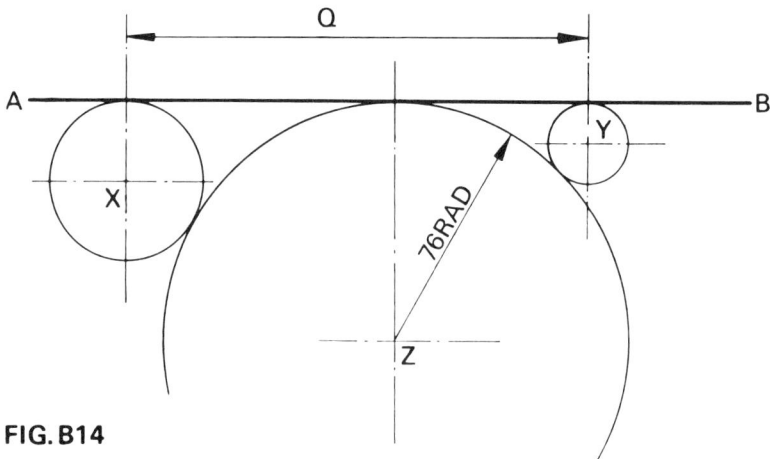

FIG. B14

3 (U.L., 1966) A straight line *AB* is a common tangent to a 50mm diameter circle *X* and a 25mm diameter circle *Y*. A circle *Z* of radius 76mm touches the line and both circles, as shown at Fig. B14.

Using a suitable construction for all points of tangency draw the figure and measure the distance *Q* between the centres of circles *X* and *Y*. State your answer to the nearest 0·2mm.

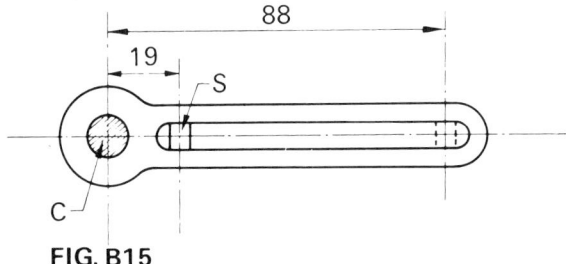

FIG. B15

4 (A.B., 1966). Fig. B15 indicates a slotted lever mounted on a shaft, centre *C*, and carrying a slider *S*. The lever makes one revolution clockwise at constant speed about *C* while the slider moves with constant speed from its minimum position from the shaft to its maximum position. Plot, full-size, the path traced out by *S*.

5 (U.O., 1969). A line *OA* rotates at a uniform speed about a fixed centre *O*. A point *B* moves at a uniform speed along *OA* starting at a point 25mm from *O*. If in one complete revolution of *OA* in an anticlockwise direction the point *B* finishes up 100mm from *O*, plot the locus of *B*. Beneath the curve print neatly its name.

C. SOLID GEOMETRY

Test 1

FIG. C1

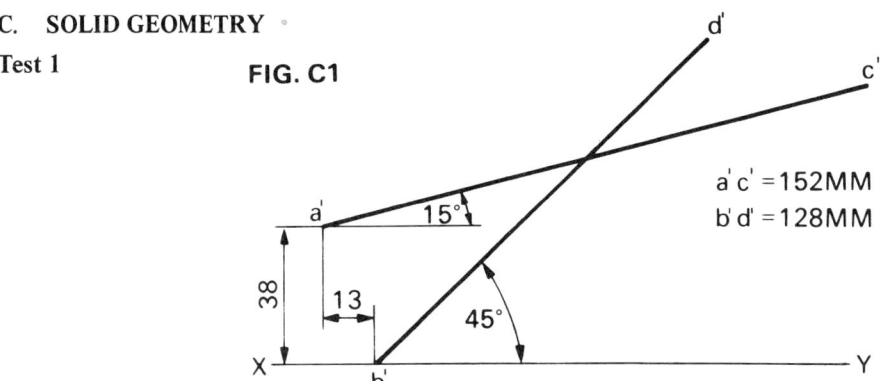

a'c' =152MM
b'd' =128MM

1 (J.M.B., 1970). Fig. C1 shows the elevations of the diagonals of the surface of a flat plate *ABCD*. Corners *A, B* and *D* are 50mm, 76mm and 13mm respectively in front of the V.P.

Draw the elevation of this surface, *ABCD*, and project its plan.

Determine, measure and state the shortest distance of the point of intersection of the diagonals from *XY*.

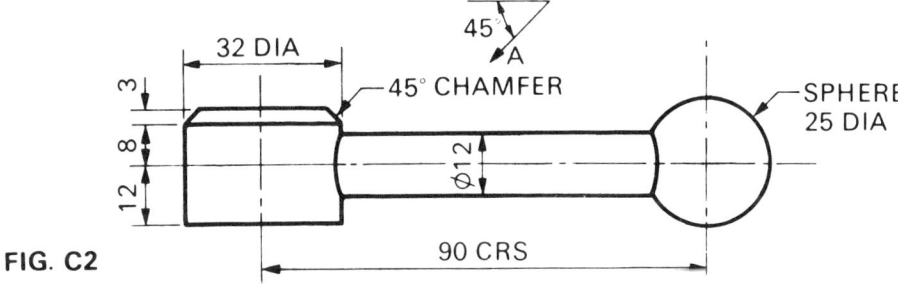

FIG. C2

2 (U.O., 1970). Fig. C2 shows details of the handle for a clamping decive. Draw the given view together with (a) a plan, (b) an elevation as seen from the left of the given view, (c) an auxiliary view when looking in the direction of arrow *A*. Use either first or third angle orthographic projection but not a mixture of both. State clearly in neat lettering which method has been used.

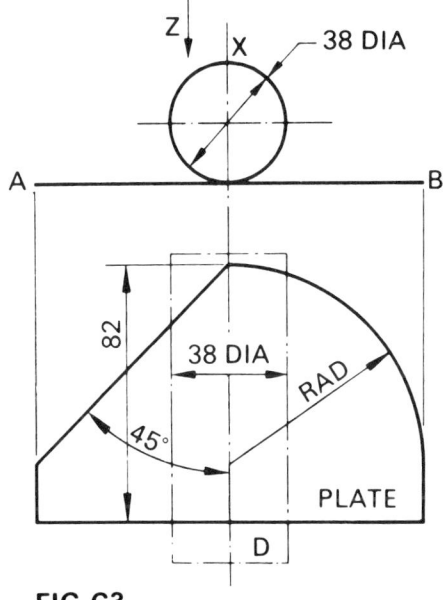

3 (U.C., 1965). The plan and elevation of a thin sheet metal plate are shown in first-angle projection in Fig. C3. A bar *D* of diameter 38mm is placed on the plate which is then tightly wrapped round the bar so that the edges *A* and *B* of the plate meet along a line at *X*. Draw a plan view of the wrapped plate when looking in the direction of arrow *Z*, assuming that the bar has been removed.

FIG.C3

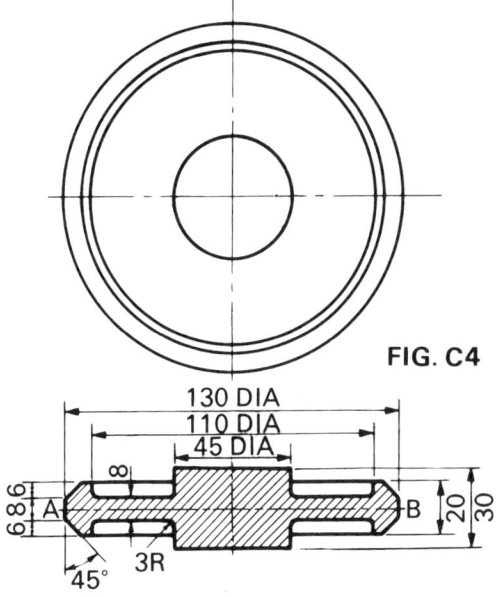

FIG. C4

4 (U.O., 1970). Details of a simple handwheel are shown in Fig. C4. Draw an isometric view of this handwheel with *AB* in the horizontal plane. Hidden detail need not be shown.

Scale: full size.

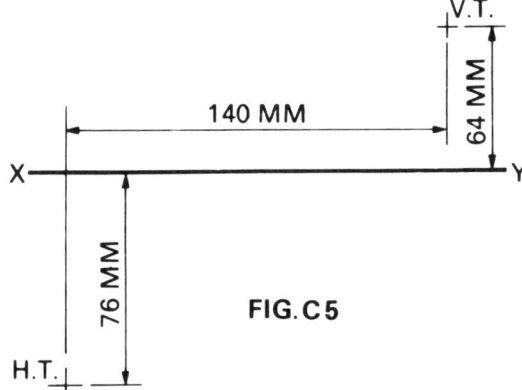

FIG.C5

Test 2

1 (J.M.B., 1970). Fig. C5 shows the traces of a straight line *AB* which has a true length of 101mm. End *A* of this line is situated 13mm above the H.P.

Draw the plan and elevation of *AB*. Determine, indicate, measure and state the angle of inclination of *AB* to the V.P.

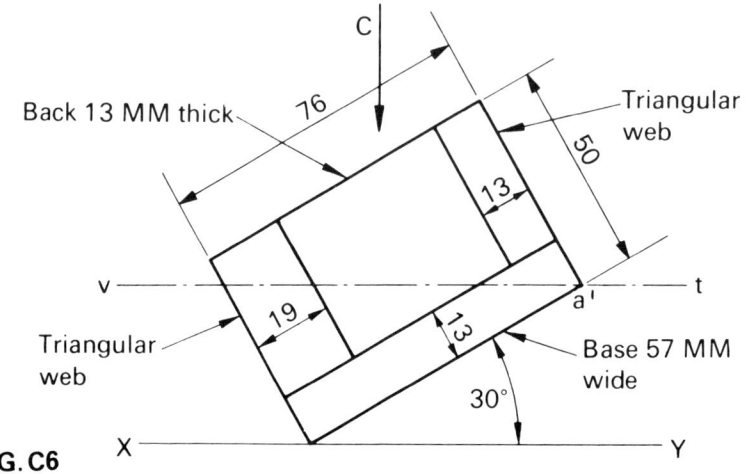

FIG. C6

2 (U.C., 1966). Fig. C6 shows the elevation of an angle bracket with two triangular webs which extend to the edge of the base. One web is 13mm thick and the other is 19mm thick as shown. Project the sectional plan on the plane *vt*, which passes through the corner *A*, when looking in the direction of arrow *C*.

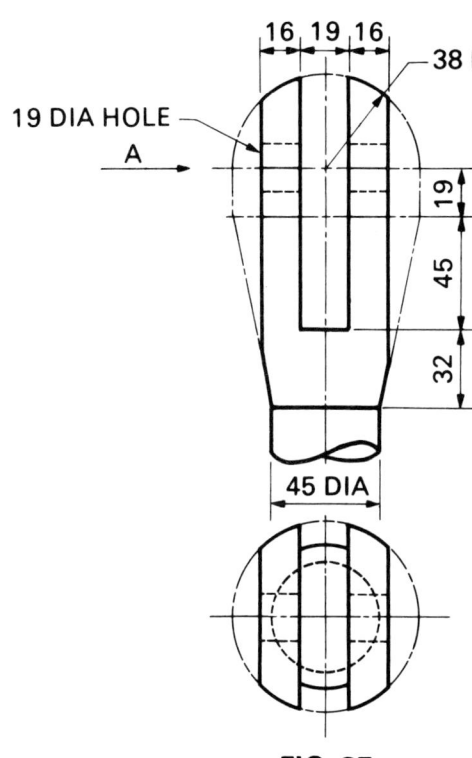

FIG. C7

3 (U.O., 1969). Two views of the fork joint of a simple coupling are shown in Fig. C7. Draw these two views together with an elevation looking in the direction of arrow A.

Scale: full size.

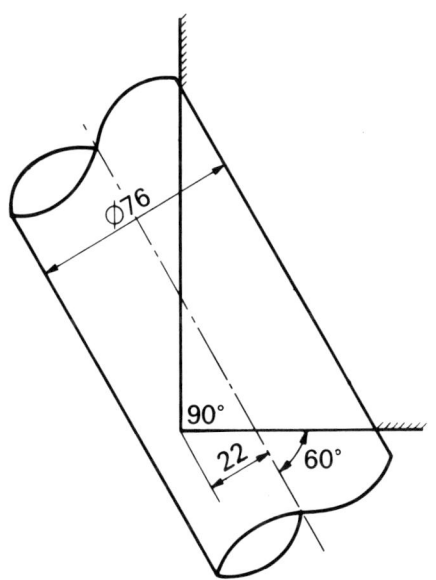

FIG. C8

4. (O. & C.B., 1970). A cylindrical pipe passes through the corner of a square tank as shown in Fig. C8. The thickness of the walls of both cylinder and tank can be ignored.

Develop, full size, one symmetrical half of the curved surface of the cylinder, showing on the development the joint line where the cylinder passes through the tank.

Draw sufficient of the length of the cylinder to include the joint.

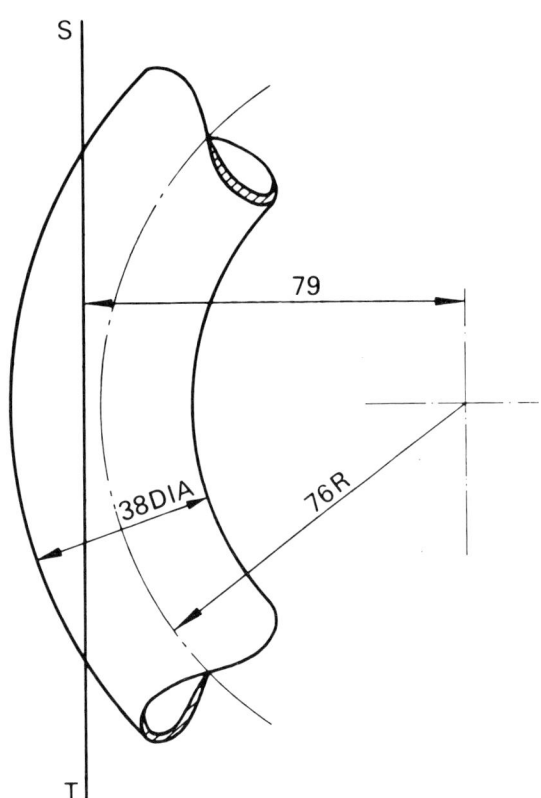

FIG. C9

5 (U.L., 1966). A pipe of 38mm diameter is bent to the arc of a circle as shown in Fig. C9. At the bend it would be obstructed by a thin plate *ST*, perpendicular to the plane of the bend.

Determine the true shape of the least hole in the plate that will enable the pipe to take up its position.

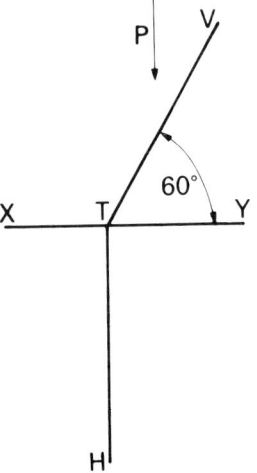

FIG. C10

Test 3

1 (O. & C.B., 1970). The traces *VTH* of a perpendicular plane are given in Fig. C10. In this plane lies an ellipse.

The end elevation of the ellipse is a circle of 64mm diameter.

Obtain:

(a) by projection, a plan of the ellipse viewed in the direction of *P*; and

(b) by any constructional method, the true shape of the ellipse.

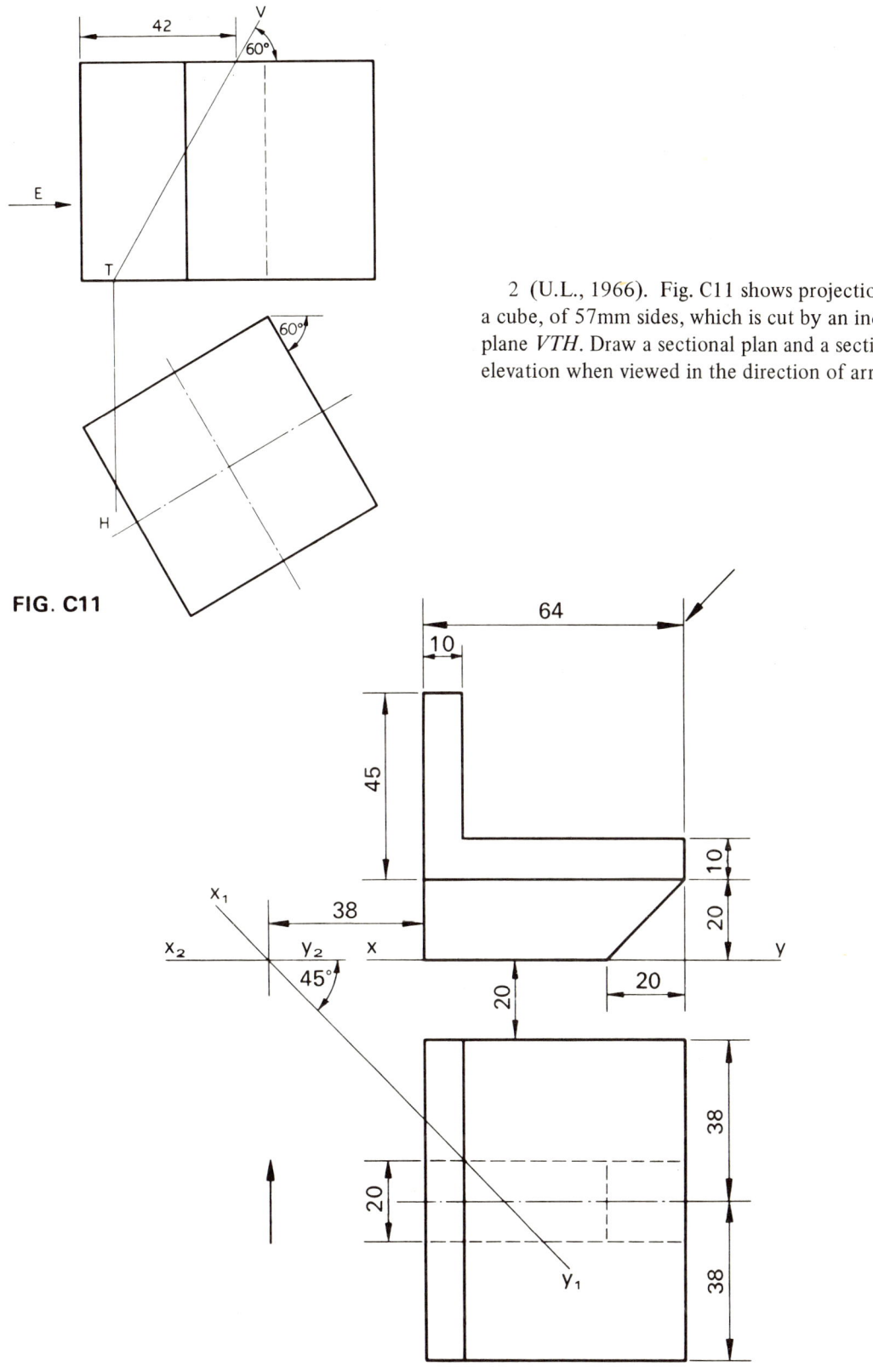

2 (U.L., 1966). Fig. C11 shows projections of a cube, of 57mm sides, which is cut by an inclined plane *VTH*. Draw a sectional plan and a sectional elevation when viewed in the direction of arrow *E*.

FIG. C11

FIG. C12

3 (A.B., 1966) The plan and elevation of a special angle bracket are shown in Fig. C12.

(*a*) Draw, full size, the given views and project an auxiliary plan on the ground line x_1y_1.

(*b*) Using the auxiliary plan in (*a*) above, project an auxiliary elevation on the ground line x_2y_2.

All hidden detail to be shown.

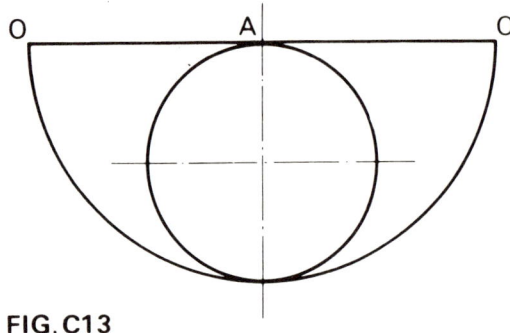

FIG. C13

4 (A.B., 1965). The development of the curved surface of a right cone is a semicircle 152mm diameter. The largest circle inscribed in the semi-circle represents a line on the curved surface, as shown in Fig. C13.

(*a*) Draw, full size, the plan and elevation of the cone when resting with its base on the horizontal plane, with the joint *AO* placed symmetrically at the rear of the cone.

(*b*) Show the curved line on the surface of the cone in both the plan and elevation in (*a*) above.

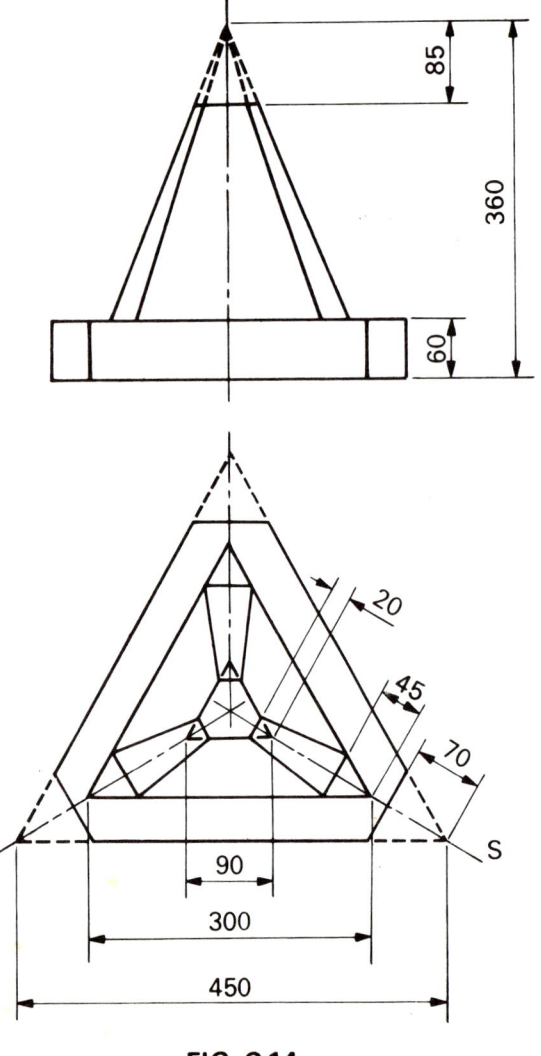

5 (O.U., Building, 1970). Fig. C14 shows the dimensioned front elevation and the plan of a stone finial; the plinth is part of a truncated equilateral triangular prism, and the vertical part is an irregular hexagonal-based truncated pyramid, the base of which is formed from the equilateral triangle of side 300mm.

To a scale of 1:2·5 draw in isometric projection the completed finial, the truncated corner, *S*, of the plinth being nearest to the view point.

FIG. C 14

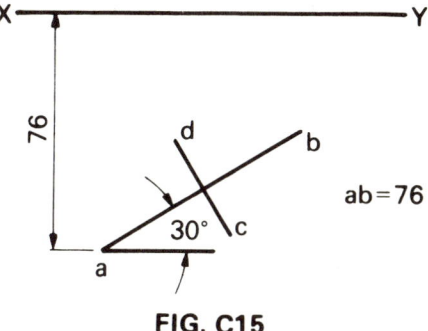

FIG. C15

6 (J.M.B., 1970). Fig. C15 shows the plan of two mutually perpendicular diameters of a 76mm diameter disc. This disc is inclined at 60° to the H.P. and *c* is in the H.P.

Draw the plan and elevation of the disc.

Determine, measure and state the shortest distance of *d* from *XY*.

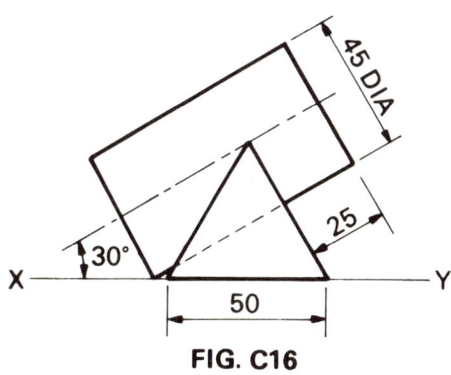

FIG. C16

Test 4

1 (J.M.B., 1970). Fig. C16 shows the elevation of an equilateral triangular prism 64mm long, shaped to receive a 45mm diameter cylinder. The axis of the cylinder lies in a plane which is parallel to the V.P. and which passes through the centre of length of the prism.

Draw the given elevation and project its plan.

Draw, also, the development of the surface of the two adjacent cut surfaces of the prism to show the cut-away required to receive the cylinder.

Indicate all hidden detail.

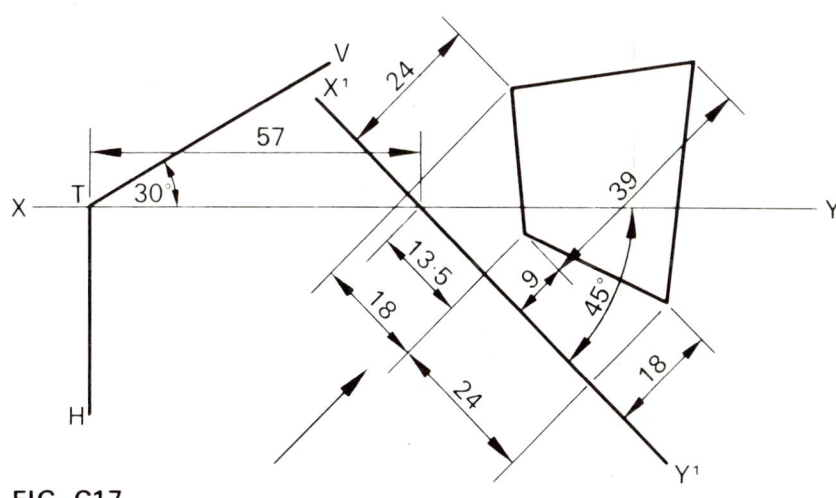

FIG. C17

2 (U.L., 1965). Fig. C17 shows an inclined plane *VTH*. This plane contains a quadrilateral which has an auxiliary elevation as shown on *X'Y'*. Construct the plan of the quadrilateral.

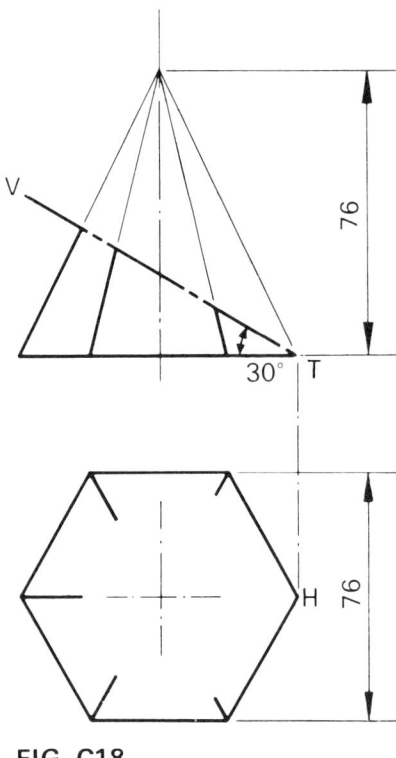

FIG. C18

3 (U.L., 1965). A hexagonal pyramid rests with its base on the H.P. and is cut by an inclined plane *VTH*, as shown at Fig. C18. Complete the plan of the lower part of the pyramid. Draw also a view showing the true shape of the section cut by the plane.

4 (A.B., 1966). A circular steel rod, 64mm diameter, lies on the horizontal plane with its axis parallel to the vertical plane. It is cut into two portions by sawing through the rod in a vertical plane inclined at 30° to the longitudinal axis of the rod. Draw, full-size, the true outline of the cut surface.

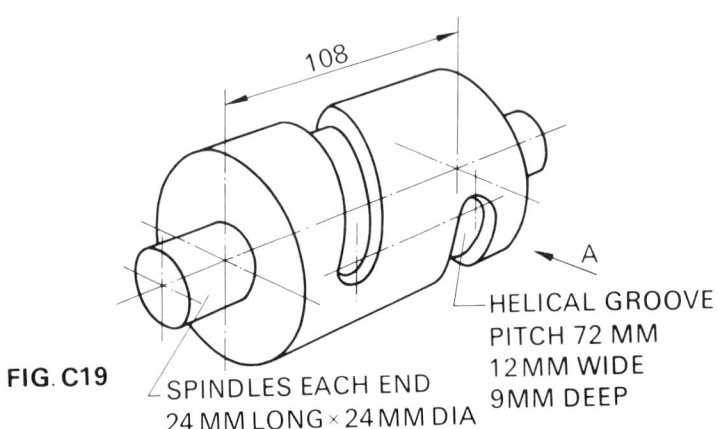

FIG. C19 SPINDLES EACH END
24 MM LONG × 24 MM DIA

HELICAL GROOVE
PITCH 72 MM
12 MM WIDE
9 MM DEEP

5 (W.J.C., 1966). Details of the operating spindle for a gear mechanism are given in Fig. C19.
Draw a complete elevation, projected in the direction of the arrow *A*, including hidden detail of the helical groove.

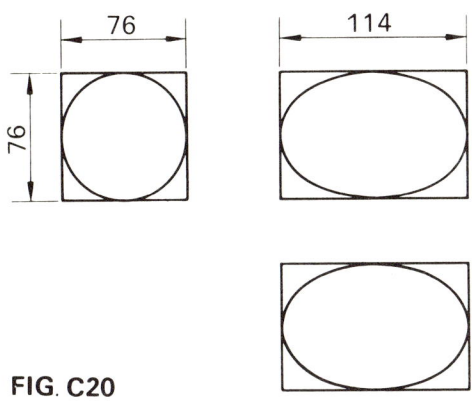

FIG. C20

6 (W.J.C., 1965). Fig. C20 shows two elevations and a plan of a square prism which has circles inscribed on its ends and ellipses on its rectangular faces.

Draw a conventional isometric projection of the prism, including the curves on its surfaces. Hidden detail is not required.

INDEX